Heterogeneous Photocatalysis Using Inorganic Semiconductor Solids

Umar Ibrahim Gaya

Heterogeneous Photocatalysis Using Inorganic Semiconductor Solids

Springer

Umar Ibrahim Gaya
Department of Pure and Industrial Chemistry
Bayero University Kano
Kano
Nigeria

ISBN 978-94-007-7774-3 ISBN 978-94-007-7775-0 (eBook)
DOI 10.1007/978-94-007-7775-0
Springer Dordrecht Heidelberg London New York

Library of Congress Control Number: 2013950803

Printed on acid-free paper

Springer is part of Springer Science+Business Media (www.springer.com)

MIX
Papier aus verantwortungsvollen Quellen
Paper from responsible sources
FSC® C105338

If you have any concerns about our products,
you can contact us on
ProductSafety@springernature.com

In case Publisher is established outside the EU,
the EU authorized representative is:
Springer Nature Customer Service Center GmbH
Europaplatz 3, 69115 Heidelberg, Germany

Printed by Libri Plureos GmbH
in Hamburg, Germany

Preface

As early as 1969, a groundbreaking study informed to launch rigorous research activity on the reactivity of photoactivated semiconductor catalysts coined photocatalysis. Staring the decades of development, research activities have come a long way in accomplishments. Currently, articles in different aspects of photocatalysis abound most scientific journals and libraries. Such advances brought forth will inform the decision to as well publish more books. The present book underscores many of the essential grounds of photocatalysis and provides updates on the scientific cornerstones, research advances, current opinions and interpretations. Basically, the book is devoted only to photocatalysis by semiconductor materials. Therefore, in major discussions other photocatalysts such as phthalocyanines will not at all be considered. Our emphasis is on treating topics essential for understanding the current drive in photocatalytic research.

We began our discussion of photocatalysis in Chapter 1 by introduction of the concepts that form the backbone of photocatalysis. The chapter covers the principles of solid state chemistry and physics underlying photocatalysis including the concept of semiconductivity, band gap calculations, defects and reactive oxidising species. Photocatalysis has an organic link with chemical kinetics and leads unifying investigations where all the branches of science cross-pollinate. Chapter 2 of the book therefore describes kinetic concepts as they apply to photocatalysis. In this chapter, the dependence of rate on reaction conditions and parameters has been detailed. Chapter 3 deals with mechanism of photocatalysis. Aspects of physical chemistry such as adsorption models, photocatalytic rates and kinetic disguises have been discussed. Chapter 4 focuses on the structure, property and activity of prototypical semiconductor photocatalysts. One poser that is invariably raised is how to extend the spectral absorption of TiO_2 to visible region to enable the effective use of visible solar spectrum. This chapter discusses the strategies to drive substantially improved photoactivity from semiconductor materials. The factors affecting performance of photocatalyst have been highlighted.

The goal of environmental photocatalytic oxidation is to eradicate air and waterborne pollutants. Chapter 5 presents a recent survey of persistent pollutants, which is aimed at helping to provide perspective or update to researchers. As most workers have used various photocatalysts, model pollutants, physical state of reactant, reactor configurations and conditions, this chapter brings to notice the useful findings. Besides environmental purification, photocatalysis finds viable applications in chemical analysis and alternative energy. The chapter also devotes discussions on recent topics in the applications of photocatalysis such as remote photocatalytic oxidation, intelligent ink indicator and dual function photocatalysis. Chapter 6 focuses on the advances that relate to organic chemistry especially as regards synthesis, product selectivity and characteristic reactions. Finally, this text utilises current terminologies and acceptable illustrations in the aspects of physical chemistry of heterogeneous photocatalysis in order to guard against misconceptions.

Readership: This book would appeal to students, and researchers in the industry and academia, who may be interested in applied physical chemistry and related fields in the mainstream chemistry tandem. I hope the book will be extended to the bookshelves of myriad libraries for extensive dissemination of knowledge.

Acknowledgements

If charity really begins at home, I will start my acknowledgements from home. I am grateful to my father Alhaji Ibrahim Abdulkadir Gaya, and my late mother Hajiya Fatima Ibrahim, for their love and care. I owe debt to my wife who endured the unpleasant period of writing this book. I express my gratitude to all brothers, sisters and friends who continuously wish me the best of academic carrier.

Many people have assisted in the success of this project in one way or another. I wish to start my acknowledgements by expressing gratitude to various workers who provided permission to reuse their illustrations in this book. To mention, I gratefully acknowledge the assistance of Professor Michael Grätzel, the father of the famous Grätzel cell and the Director of Photonics and Interfaces Laboratory, Ecole Polytechnique Federale de Lausanne, Switzerland. Accordingly, I express gratitude to Professor Naoto Tamai, Department of Chemistry, School of Science and Technology, Kwansei Gakuin University, Japan. I owe heaps of thanks to Professor Hadis Morkoç of the Department of Electrical Engineering and Physics Department, Virginia Commonwealth University, Richmond, USA. Similarly, I thankfully acknowledge the support of Prof. Russell Howe, the Head of the Department of Chemistry, University of Aberdeen, UK and Professor Shizang Qiao of the ARC Centre of Excellence for Functional Nanomaterials, University of Queensland, Australia and Prof. Dr. H.G. Yang of the Key Laboratory for Ultrafine Materials, School of Materials Science and Engineering, East China University of Science and Technology, China. I also wish to acknowledge the American Chemical Society, Elsevier, Royal Society of Chemistry, and Taylor and Francis Group for permission to reuse copyright materials.

I am quite indebted to all my colleagues in Bayero University Kano, and in other places I dwelled for different academic pursuits for all fruitful discussions we had before and while writing this book. Special thanks to Dr M.A. Kurawa, the present Head of Pure and Industrial Chemistry Department, Bayero University Kano, as well as other members of the department. I thank Dr Abdul Halim Abdullah and Professor Zulkarnain Zainal, the former Heads of Department of Chemistry, and Professor Taufiq Yun Hin and Professor Mohd Zobir Hussein, the former and the current Heads of the Putra Centre of Excellence for Catalysis Science and Technology (PutraCat), for their useful advice during my days as student at Universiti Putra Malaysia. I wish to express gratitude to Professor Detlef Bahnemann of Institut fur Technische Chemie, Gottfried Wilhelm Leibniz, University of Hannover, Germany, for sharing with me sometimes earlier a few of his publications on doped titania materials and on the preparation of brookite. Finally, I gratefully acknowledge the time and energy allocated by those who participated in the blind review of this book, and the staff of Springer who interfaced with me.

Umar Ibrahim Gaya

Contents

Acronyms and Symbols

AC	activated complex
A_s	specific surface area of reactor
BET	Brunauer-Emmett-Teller
CB	conduction band
CVD	chemical vapour deposition
D	diffusion coefficient of charge carrier (or effective diffusivity)
d	size of the semiconductor particle's cross-section
DRS	diffuse reflectance spectroscopy
E	applied electric field
E^*	apparent activation energy
E_F	Fermi-level energy
E_g	band gap energy
E_i	ionisation energy
E^o	standard oxidation potential
EPR	electron paramagnetic resonance
FID	flame ionisation detector
FWHM	full width at half maximum
g	electron g-factor
GC/FID	gas chromatography/flame ionisation detector
GC-MS	gas chromatography-mass spectroscopy
H	thickness of semiconductor film
HPLC	high performance liquid chromatography
IR	infrared
J	current density
k	rate constant
K	Boltzmann constant
k'	pseudo order rate constant
K_{ads}	adsorption constant
k_m	fluid to surface mass transfer
L	crystallite particle size
LDPE	low density polyethylene
m	evaporation rate

NMR	nuclear magnetic resonance
P.E.	potential energy
pKa_1	negative log of first acid dissociation constant
pKa_2	negative log of second acid dissociation constant
PVD	physical vapour deposition
R	molar gas constant
r	radius
r	rate of reaction
R	size of agglomerate
RF	radiofrequency
ROS	reactive oxidising species
SAED	selected area electron diffraction
SC	semiconductor
TEM	transmission electron microscopy
TOC	total organic carbon
TON	turn over number
UV	ultraviolet
VB	valence band
XPS	X-ray photoelectron spectroscopy
XRD	X-ray diffraction
α	absorption coefficient
ε	permittivity of space
η	optimal catalyst domain
θ	diffraction angle
$\mathrm{\theta}$	surface coverage
μ	viscosity
μ_B	Bohr's magneton
ν	frequency (of light or light absorption)
ξ_r	relative photonic efficiency
ρ_{sc}	density of semiconductor photocatalyst
σ	conductivity or surface charge density
τ	fractional lifetime or half-life
ϕ	probability of light absorption to produce charge carriers
Φ	quantum yield (or thiele modulus)
ϕ_{eff}	photocurrent efficiency
φ_L	light harvesting efficiency
ω	angular velocity

Chapter 1
Principles of Heterogeneous Photocatalysis

Abstract At the turn of 1970s, the photo-response of solid materials became a constant subject of intensive research for possible application in modern technologies. Heterogeneous photocatalysis has evolved in a unique way giving rise to various applications for alternative energy, organic synthesis and environmental treatment. Semiconducting compounds have thrived as the most suitable materials for these applications. Towards achieving the maximum potential benefits of semiconductor photocatalysis and to make remarkable impact, the fundamental concept as well as application has been extensively and rigorouslyexplored. We therefore detail herein the basic formalism of semiconductor photocatalysis from band models through the generation of oxidants to the light harvesting processes. The key imperfections in semiconductor oxides and their consequences on the extrinsic semiconductivity are outlined. The advances being made regarding the role of oxidising species in photocatalytic degradation are rationally discussed. Finally, the reader is referred to some 'must read' articles in heterogeneous photocatalysis by semiconductor solids.

1.1 Background Sketch

The interaction of light with molecules and the resulting processes are currently of major interest in the development of various chemical technologies. Various types of semiconductor catalysts are known to be photosensitive, allowing activation by light photons. Phoactivated molecule may initiate or influence the rate of chemical reaction by energy or charge carrier transfer. Such molecules are known as photocatalysts and they form the backbone of heterogeneous photocatalysis. Heterogeneous photocatalysis has gained acceptance as a viable alternative to traditional biological, chemical and physical decontamination technologies owing to its striking attributes for exploitation in environmental context, which include non-selectivity, non-toxicity, total removal of pollutant and by-products, and cost-effectiveness (Tai and Jiang 2005; Doong et al. 2000; INERIS 2005; Muruganandham et al. 2006; Saien and Soleymani 2007).

The world has constantly been witnessing advances in the manufacture and use of synthetic chemicals. Regrettably, a high number of these chemical substances are

U. I. Gaya, *Heterogeneous Photocatalysis Using Inorganic Semiconductor Solids,* DOI 10.1007/978-94-007-7775-0_1,

known to have severity of toxicity to both plants and animals (Pelizetti et al. 1990; Abramović et al. 2004). Certainly, the environmental implication of such hazardous substances cannot be tolerated. Against this background, environmental agencies have periodically come up with legislations to guide the use of synthetic chemicals and release of waste (David 1995; Saxena and Jotshi 1996; Carp et al. 2004). In spite of these stringent regulatory measures however, pollution has become widespread owing to varied nature of wastes, contributory negligence to control the use of waste chemical products or poor efficiency of treatment technologies.

Several environmental pollutants have become prominent, among which dyes and halophenols stand in the forefront owing to their widespread uses. About 1–20 % of the world production of dyes is loss during the dying process (Kabra et al. 2004; Augugliaro et al. 2006). The chemical nature and stability of these synthetic dyes spells serious threat to the environment. Similarly, since 1930s halophenols have been used as chemical intermediaries, wood preservatives, miticides, pesticides, herbicides and as defoliants (Herrmann 1999; Lewis 2002). About 85 % of these organic wastes may be found in surface water (Pohanish and Green 1996). Well aware of the implications of hazardous substances, scientists have continually developed methods which could serve to remove these important pollutants from the environment. Several conventional methods therefore exist which include adsorption and thermal oxidation. In retrospect, many of these traditional treatment methods have been identified with various disadvantages such as non-destructiveness (US EPA 2006), poor rate of removal at higher concentration (EPA 1998), high cost or ineffectiveness at low pollutant level (US EPA/OSHA 2000).

Photocatalysis is a chemical jargon conventionally defined as the acceleration of the rate of a chemical reaction, induced by the absorption of light by a catalyst or co-existing molecule. This may be the most widely accepted definition of photocatalysis as it encompasses all aspects of the field including photosensitisation (Serpone and Salinaro 1999; Braslavsky 2007). Photosensitisation is the light-induced alteration of a molecular entity as a result of photoabsorption by another molecule (the photosensitizer). Although the initial molecule does not absorb light, this phenomenon can still be classed under photocatalysis provided that it exhibits catalytic nature. That is to say the photosensitizer must bring about a positive change in quantum efficiency (Ohtani 2008). Generally, a photocatalyst is a material that can induce reaction upon efficient light absorption. The number of available photocatalysts is high, and these photocatalysts may be categorised into semiconductor-based, composites, complex oxides, polyoxometallates and complex organic or organometallic compounds such as porphyrin, pthalocyanine, metalloporphyrins and metallophthalocyonines.

Phthalocyanine is synthetic macrocyclic derivative of porphyrin having an extended π-electron system (**1**). When phthalocyanine (**2**) is complexed by a metal (with the metal in the centre of the macrocycle) the resulting derivative is called metallophthalocyanine. Several transition metal phthalocyanine complexes are known and they tend to be more resistant than porphyrin itself. Phthalocyanine and metallophthalocyanines possess photocatalytic activity for a wide variety of reactions. These substances qualify further as heterogeneous photocatalysts for organic pollutant removal owing to their poor solubility in both organic solvents and water. From synthesis point of view, the preparation of metallopthalocyanines is quite simple. For example, zinc phthalocyanine can be facilely prepared by condensation of phthalic anhydride in presence of urea.

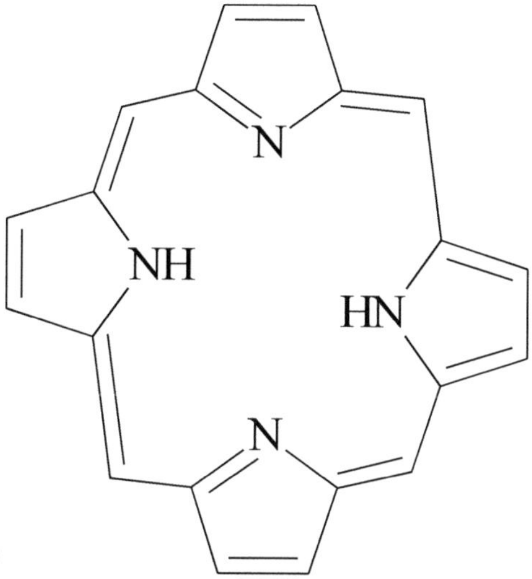

(1)

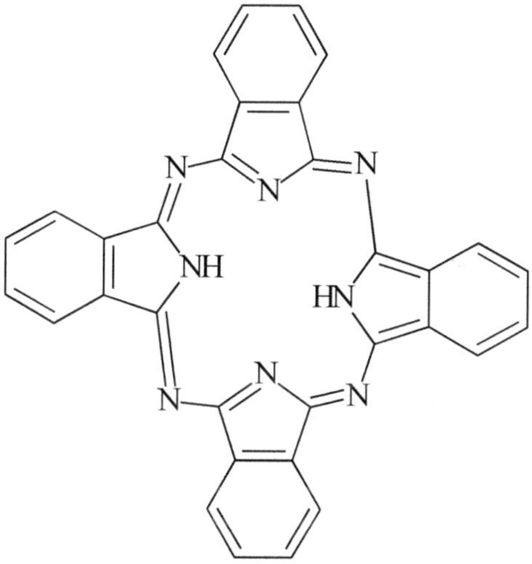

(2)

The mechanism by which giant organic or organometallic molecules serve as photocatalysts is simply photosensitisation. Metallophthalocyanine (MPc) (3) can be excited to singlet state ($^1MPc^*$) by irradiation, which through intersystem crossing moves to the triplet excited state ($^3MPc^*$). The triplet state then interacts with ground (triplet) state molecular oxygen ($^3O_2$) to generate excited singlet oxygen ($^1O_2$) for the oxidation of organic molecules as shown in the reactions (1.1) to (1.4).

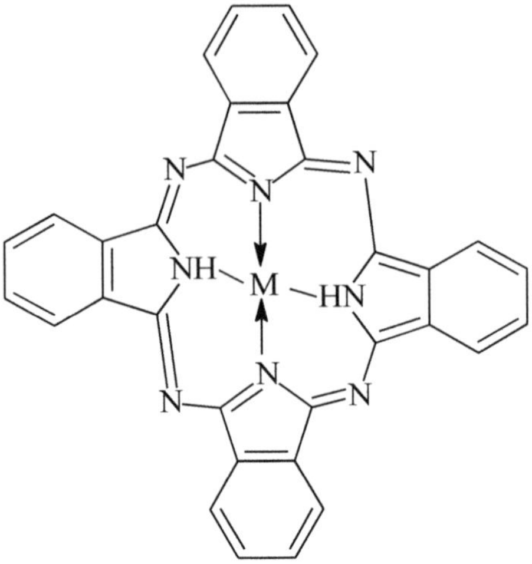

(3)

$$MPc \xrightarrow{hv} {}^{1}MPc^{*} \tag{1.1}$$

$${}^{1}MPc^{*} \xrightarrow{\text{intersystem crossing}} {}^{3}MPc^{*} \tag{1.2}$$

$${}^{3}MPc^{*} + {}^{3}O_2 \rightarrow MPc + {}^{1}O_2 \tag{1.3}$$

$$\text{Substrate} + {}^{1}O_2 \rightarrow \text{oxidation products} \tag{1.4}$$

The use of phthalocyanines as photocatalysts for environmental decontamination suffers some drawbacks. Even though several metallophthalocyanines can be listed (with central metal as Zn, Al, Si, Co, Ni, Cu or Ti) only the aluminium, silicon and zinc -based counterparts were sufficiently photocatalytically active (Kluson et al. 2007). Several workers have therefore attempted the enhancement of photoactivity of the metallophthalocyanines by the use of support materials in the case such as polydivinyl benzene (Wu et al. 2007), TiO_2 and Al_2O_3 (Iliev 2002). Besides, phthalocyanines and metallophthalocyanineslo present real perspectives as photocatalysts because their relative stability depends on their molecular structure and whether supported or not (Mele 2003).

Polyoxometallates have been investigated as homogeneous photocatalysts for photocatalytic degradation of organic substances. Polyoxometallates (POMs) are molybdates and tungstates of the Keggin structure (composing of Keggin anion $[XMo_{14}O_{24}]^{n-}$ where X = P, Si, Fe, H_2), Dawson type structure (composing of $[P_2W_{18}O_6]^6$) or similar structure. So, we can have a polyoxometallate from Keggin anions such as $[SiMo_{14}O_{24}]^{4-}$, $[PMo_{14}O_{24}]^{3-}$ while cations may be H^+, K^+, Na^+ or $NH_4{}^+$. The molecular excited states of the molecules in the reactivity system (the lowest unoccupied molecular orbital, LUMO and the highest occupied molecular orbitals, HOMO) initiate an electron tranfer process via which efficient photodegradation may take place. For instance, the photodegradation of 2-mercaptobenzothiazole has been studied using a polyoxotungstate ($Na_4W_{10}O_{32}$) (Allaoui et al. 2010). The POM increased the rate of 2-mercaptobenzothiazole relative to photolysis by 6 times. The disadvantage of the use of POMs as photocatalysts stems from the fact that they are highly soluble in polar solvents.

Relatively, inorganic semiconductors have been recognised as the most successful photocatalysts for various forms of applications. These substances occur in the form of oxides, sulphides, composites and complex oxides. Some notable examples of these semiconductor solids used in environmental heterogeneous photocatalysis include TiO_2, ZnO, CdS, WO_3, SnO_2, ZnS, CdTe, CdTe, α-Fe_2O_3, $AgNbO_3$ and $SrTiO_3$. These compounds demonstrated rich photochemical properties for harvest. For instance, CdS can absorb most part of visible light owing to its short band gap (2.42 eV) which makes it a potential photocatalyst for solar cells and photocatalysis. However, this material is not very stable under photocatalytic condition. Similarly, ZnO exhibits several important characteristics such as piezoelectric properties and extended spectral absorption. This photocatalyst has sometimes surpassed TiO_2 in terms of photocatalytic efficiency but a problem associated with it is propensity for photodissolution.

By far, TiO_2 and TiO_2-based materials have been identified as the most economically attractive and authoritative photocatalysts owing to properties like non-toxicity, cost-effectiveness, ready availability, chemostability, reusability, outstanding electronic and optical characteristics. Consequently, there are various purposes for which this inorganic semiconductor can be used. Worth mentioning here are environmental cleaning, solar-energy conversion, photodisinfection and photokilling of cancer cells. In the environmental cleaning context, TiO_2 is used in self-cleaning glasses and in the mineralisation of aqueous or air borne pollutants. The disinfection of water and air from microorganisms can also be achieved by TiO_2 photocatalysis. A good example of the use of titania in energy applications is the dye-sensitised photovoltaic cells. This cell consists of TiO_2 duedue to its superior applicability which includes resistance to photocorrosion (Desilvestro et al. 1985; Argazzi and Bignozzi 1995; Tachibana et al. 1996). At the moment, photocatalysis has become a point of focus where innumerable disciplines converge.

Semiconductor-mediated heterogeneous photocatalysis has appeared in various forms and has therefore attracted a high number of reports dealing with fundamental studies, integration and application (Dillert et al. 1999). In fact, a high number of patents and applications on photocatalytic technology have appeared proving its unrivalled power as an environmental treatment tool (Blake 1999). Decidedly, these projects need to be guided by what we furnish herein as the waterfall model of photocatalyticdevelopment (Fig. 1.1). Arguably, every successful photocatalytic investigation ought to go through the cyclic model.

In the first stage of the waterfall model new ideas and proposals are considered. This stage can therefore be referred to asthe feasibility stage. The next stage of the lifecycle involves laboratory analysis in order to arrive at optimised parameter conditions for successful laboratory operation. Subsequently, a design of a pilot reactor may be made and performance tests carried out. Finally, a large-scale facility

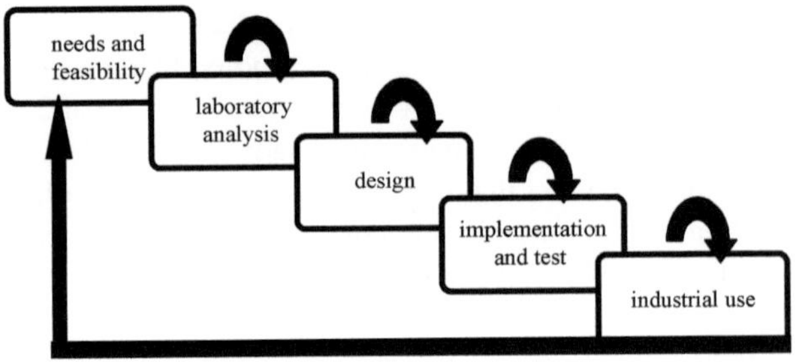

Fig. 1.1 Waterfall model for the development of photocatalytic application

can be generated. Feedbacks arising from these large-scale applications are tackled by re-cycling through the waterfall cycle.

1.2 Evolution of Semiconductor Photocatalysis

From historic point of view, photocatalysis has been an old phenomenon and its existence is apparent from common observations. Ideas on photocatalysis can be visualised from the fading of painted walls upon long-term exposure to sunlight. Fundamentally, this illustrious phenomenon takes place as a result of the photo-oxidative reactions the paint undergoes in present of the TiO_2 used as additive. Even though studies on the photoactivity of semiconductor oxides for the oxidation of substances can roughly be traced to early twentieth century, the question of where it first originated may continue to be unanswerable (Hashimoto et al. 2005). In 1921, Renz (1921) found for the first time that titanium oxide, as well as some other oxides investigated, formed dark reduction products when irradiated in the presence of a suitable organic acceptor. The utilisation of TiO_2 for the photo-decolourisation of dye was studied way back 1938 by Goodeeve and Kitchener (1938), and the oxidation of organic solvents in 1950s, by Kato and Mashio (1956). Later on Morrison and Freund (1967) studied the photocatalysed oxidation of aqueous phase formic acid in presence of ZnO which yields CO_2 and H_2O_2. With the report of CO_2 and H_2O_2, the study perhaps places ZnO the first semiconductor to be used in photocatalytic mineralisation.

It was not until 1969 when Fujishima and Honda (1972) made a groundbreaking discovery of the photoelectrolysis of water molecule using an *n*-type rutile TiO_2 anode against platinum cathode that photocatalysis came into much limelight and prompted the beginning of fervent research in the field of photon energy conversion as an alternative energy source, through 1970s to early 1980s. Subsequently, deriving useful reactions by this great phenomenon became the business of the day. Today, several applications of photocatalysis can be clearly distinguished even domestically.

1.3 Photophysics and Photochemistry

At molecular level, the interaction of the semiconductor with light is considered more or less in unified way by all workers. The core concepts of heterogeneous photocatalysis are treated in the framework of photophysics and photochemistry. However, the resulting processes and applications can be viewed from various standpoints. Many semiconductor photocatalysts would naturally be thought of to be visible-light responsive owing to their possession of colour. For instance, WO_3 and CdS have rich yellow colour are good visible light photocatalysts. The first explanation that will come to mind for their photocatalytic activity is therefore the excitation of electrons. Basically, when light is absorbed by the semiconductor photocatalyst electronic excitations take place, which can be exploited in chemical or electrical work. The nature of application of these chemical excitations may vary depending on discipline. As a result, potpourri of processes and applications has been established using various photoactive semiconductor materials.

Materials have often been classified on basis of their ability to conduct electricity mainly into three classes: conductor, insulator and semiconductor. In a conductor electrons are free to move from one site to another whereas in an insulator there is no movement of electrons at all. Semiconductors are characterised by intermediate electronic behaviour. Semiconductors have more readily understood electronic structure than conductors, usually explained by the 'band model' (Hagen 2006).

1.3.1 Band Gap Model

The atoms or molecules of a solid are grouped together in certain order as well as with their energy levels. The energy levels are grouped in form of bands. The band gap theory proposed two bands for a crystalline solid: the conduction band CB (being the lowest unoccupied molecular orbitals, LUMO; consisting antibonding orbitals) and the valence band VB (being the highest occupied molecular orbitals, HOMO; made up of bonding orbitals) which are separated by forbidden zone called the band gap. In insulators the conduction band and the valence band are separated by a large band gap. In metallic conductors however, there is apparently absence of gap between the CB and the VB owing to overlap of molecular orbitals. Semiconductors have relatively intermediate band gap and can be distinguished from insulator by band gap size. A standard of 3 eV has usually been regarded as the highest limit for the classification of semiconductors with less consistency as many semiconductors such as ZnO and TiO_2 possess larger band gaps than the standard. Perhaps it will be more appropriate to stick to the definition of semiconductor based on conductivity.

In terms of conductivity, semiconductor can be defined as a solid whose electrical conductivity (or resistivity) is intermediate between that of a conductor and an insulator (Carmen 2009). Conveniently therefore, the form of Ohm's law shown in

Eq. (1.5) can be used to distinguish between conductor, semiconductor and insulator (Richard 1995).

$$J = \sigma \xi \tag{1.5}$$

where J is the current density (A cm^{-1}), σ is the conductivity (mho cm^{-1}) and ξ is the electric field applied. Conductors and semiconductors exhibit conductivities within ranges of 10^6–10^4 and 10^{-9}–10^3 respectively, while insulators dart in the ranges of $10^6 - 10^4$ and $10^3 - 10^{-9}$ mho cm^{-1} respectively, while insulators dart in the range of $10^{-9} - 10^{20}$ mho cm-1. This classification places the conductivity of semiconductor between that of conductor and insulator with specified limits. It may be important to mention that although semiconductors do not conduct electricity at low temperature they do so at higher temperatures (Hagen 2006; Willie 2010).

1.3.2 Semiconductivity and Energetics

Semiconductivity is exhibited by ionic compounds which do not contain stoichiometric amount of metal and non-metal such as ZnO and SnO_2, or by insulators that are slightly contaminated (Castellan 1983). Three types of semiconductors may be distinguished (Hagen 2006): *i*- (or intrinsic), *n*- and *p*- (or extrinsic) semiconductors. An *i*-semiconductor is a perfect crystal semiconductor containing no impurity while *n*- and *p*- semiconductors contain impurity in their crystal lattice. Impurities in this context may also be referred to as dopants. Intrinsic semiconductors have no impurity type of imperfection but may have defects which do not affect the overall composition of the semiconductor. Defects of such type are also known as intrinsic defects. In intrinsic semiconductor, conduction occurs by thermal excitation of electrons from valence band to conduction band.

The impurities responsible for extrinsic semiconductivity can either be electron donors or electron acceptors. The introduction of an electron donor into the forbidden band of a semiconductor makes it easier for an electron to be added to the conduction band (CB) without corresponding generation of hole (h^+_{VB}) in the valence band (VB). Consequently, there is stronger presence of electrons compared to the positive holes and the resulting semiconductor is termed *n*-type semiconductor. Conversely, if an electron acceptor is incorporated into the crystal lattice of a semiconductor it becomes less easy for an electron to be added to the CB. In this case, the majority carriers are the positive holes while the minority carriers are the electrons. The resulting material is called *p*-semiconductor. In either case, the ionisation energy (E_{ion}) to eject electron from the valence band is low and can even be supplied thermally.

In the forbidden zone there exists an energy level known as Fermi, at which the probability of electron occupancy $f(E)$ is 0.5. The Fermi level lies above the donor (dopant) level of *n*-semiconductor, or below the acceptor (dopant) level for a *p*-type semiconductor. Therefore, Fermi level is closer to conduction band in *n*-type semiconductorsand closer to the valence band for *p*-semiconductor. A comparative

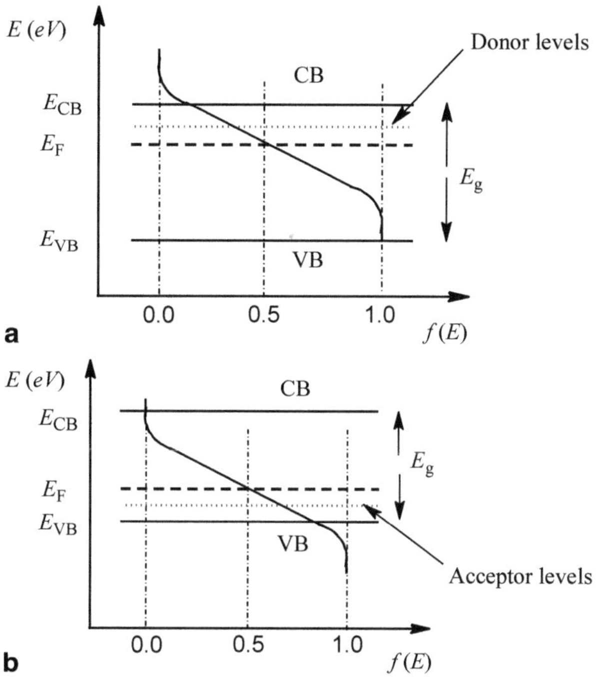

Fig. 1.2 Schematic diagrams showing the electron energy levels in extrinsic semiconductors: (**a**) *n*-semiconductor (**b**) *p*-semiconductor. E_{CB}, E_{VB}, E_g and E_F represent energy edges of conduction band, valence band, band gap and Fermi level

illustration of energy edges of *n*- and *p*- semiconductors is shown in Fig. 1.2. Unlike insulators, it is easy to fine tune the properties of these semiconductors to suit industrially important applications such as piezoelectricity, chemical sensors, varistors, solar cells, light emitters and as catalysts (Ozawa et al. 2005).

Semiconductor photocatalysts are a priori prepared in doped or extrinsic semiconductor form. The commonest examples of semiconductor photocatalysts such as TiO_2 and ZnO are *n*-type semiconductors. Generally, the electronic mechanism for the formation of the *n*-type or *p*-type is chemically represented by defect reaction. The excess cations or anions resulting from doping may be situated in part of the semiconductor lattice known as impurity defect. Defects are formed via physical processes. The defect processes for the formation of *n*-type or *p*-type semiconductor cancan be given by Eq. (1.6) and (1.7).

$$n\text{-type: SC} + \delta\text{D} \rightleftharpoons \text{D-SC} + \delta\text{e}^- \tag{1.6}$$

$$p\text{-type: SC} + \delta\text{A} \rightleftharpoons \text{A-SC} + \delta\text{h}^- \tag{1.7}$$

where SC is the semiconductor, δD and δA respresent small excess of donor or acceptor. The defect processes for the formation of *n*-type and *p*-type semiconductor can be more clearly illustrated using band gap model in Fig. 1.3.

It would be seen from the Fig. 1.3 that in the formation of *n*-type semiconductor the impurity the donor supplies electrons to the conduction band. Conversely, the formation of *p*-type semiconductor involves the acceptance of electrons from the

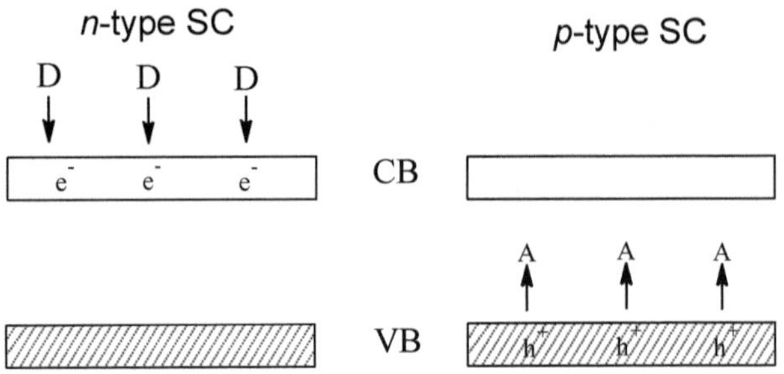

Fig. 1.3 A schematic diagram showing the formation of *n*-type and *p*-type semiconductors

valence band. For this reason *n*-type semiconductors are referred to as electron conductors while *p*-type semiconductors are referred to as hole conductors.

1.3.3 Defects in Semiconductor Solids

To make more sense of the concept of defects, we may need to touch the solid state chemistry regarding semiconductor defects and the thermodynamics of their formation. In our discussions, we shall adopt the use of Kræger notations for denoting defect terms. Native defects are associated with non-stoichiometric photocatalysts. In the case of semiconductor oxides such as TiO_2, oxygen vacancies (V_O) and metal vacancies (V_M) may prevail, which are native defects that easily form during the preparation of these materials. The term vacancy refers to a missing atom. Oxygen vacancies play several roles in the bulk TiO_2 and have potential heavy effects on material properties. For example, the higher reactivity of anatase phase of TiO_2 compared to the rutile may be traced to the difference in the locations of oxygen vacancies in either of the polymorphs (Mattioli et al. 2008). Other forms of defects may involve interstitial metals (V_{Mi}) which are formed in the interstices rather than in the electronic structure of the semiconductor photocatalyst.

Generally speaking, there are many defects in inorganic solids although only five categories of these defects have been experimentally observed: Schottky, Frenkel, anti-Frenkel, anti-structure and vacancy structure (Ropp 2003). These defects can further be divided into two types: intrinsic and extrinsic defect. Of specific interest to this text are the intrinsic defects which comprise of, for example, Schottky and Frenkel type defects associated with TiO_2 and ZnO respectively. The Schottky and Frenkel defects are stoichiometric, that is to say that they do not change the overall composition. In Schottky, defect vacancies are situated in the lattice while the Frenkel defect is formed by the movement of an atom or ion into the interstitial position (Smart and Moore 2005).

Both Schottky and Frenkel intrinsic defects can be described based on equilibria and thermodynamics. Majority of the common semiconductor photocatalysts are binary compounds of formula MX_S, where S is a small number usually from 1 to 4. For a given MX_S lattice, Schottky defect pair (Ti-O for example) may be formed via

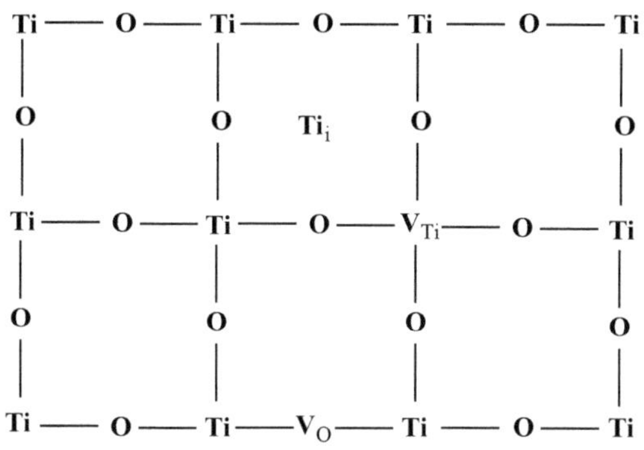

Fig. 1.4 Schematic diagram illustrating interstitial defect, oxygen and Ti vacancies in TiO_2 lattice

the following dissociation equilibria:

$$Mx_s \rightleftharpoons v_m + v_x \tag{1.8}$$

(defect pair)

Such transformation is described accordingly by corresponding defect reaction displaying the change from perfectness (Eq. (1.9)).

$$0 \rightleftharpoons V_M + SV_x + \alpha V_i \tag{1.9}$$

Where V_M is the cation site vacancy (associated with M), V_X is the anion site vacancy (associated with X) and V_i is the interstitial vacancy and αV_i represents interstices globally. The defects associated with anion site, interstitial vacancy and metal vacancy in TiO_2 are exemplified by Fig. 1.4.

In ZnO, a number of intrinsic defects have been identified with different ionisation energies: oxygen vacancy (V_O), Zn vacancy (V_{Zn}), Zn interstitial (Zn_i), O interstitial (O_i) and antisite Zn (Zn_O) (Fan et al. 2013). However, Zn interstitials and oxygen vacancies are known to be the predominant ionic defect types which in other words means the prevalence of Frenkel defects. In most cases of defect formation, if Schottky defect predominates in associated defects of a semiconductor Frenkel or other defects are usually absent (Ropp 2003). This is because the intrinsic defect that predominates is usually the one with lower endothermicity. In other words, the defect with lower ΔH^o value is more likely to predominate.

According to the second law of thermodynamics all solids must have defect. Such defects are sometimes associated with thermodynamically reversible reactions. Reactions for which equilibrium is readily established may be termed thermodynamically reversible reactions. Titanium dioxide exhibits defect disorders involving a number of intrinsic point defects which are thermodynamically reversible, as well as extrinsic defects (Nowotny et al. 2008). In the case of such a semiconductor, transition state theory may therefore apply. Consider the reversible formation of 'pre-defect' B* from A:

$$A \rightleftharpoons B^* \tag{1.10}$$

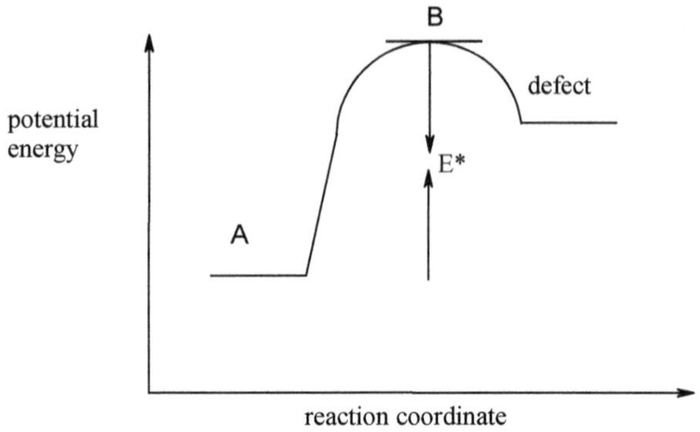

Fig. 1.5 Schematic uphill potential energy profile for defect formation

From transition state theory A is in quasi-equilibrium with B*. Let the concentration of A and B* be [A] and [B*] respectively. Let the rate constant for the formation of B* and A be k_f and k_b respectively.

The equilibrium constant (K_d) of defect formation may be derived from the overall rate of defect reaction shown by Eq. (1.11).

$$\frac{d\,[B^*]}{dt} = k_1[A] - k_b[B^*] \tag{1.11}$$

At equilibrium $d[B^*]/dt = 0$

$$\therefore \quad k_1[A] = k_b[B^*] \tag{1.12}$$

But $K_d = \frac{k_f}{k_b}$

$$\Rightarrow \quad K_d = \frac{[B]}{[A]} \tag{1.13}$$

By making [B*] the subject of the formula we obtain

$$[B^*] = K_d[A] \tag{1.14}$$

The defect formation is usually thermodynamically uphill ($\Delta G^o > 0$) which implies that the defect formed is usually having higher energy than the perfect material. In addition, the rate of the defect reaction depends on the concentration of the semiconductor having the right energy to form defects which is indicated by the concentration passing through the saddle of the potential energy with reaction progress (Fig. 1.5).

According to transition theory the rate of the formation of the transition defect r, can be given by

$$r\alpha[B^*]x\nu \tag{1.15}$$

or

$$r = \nu K_d[A] \tag{1.16}$$

where v is the frequency of formation of the transition defect, and is related to the vibrational energy required for defect formation by $E_{vib} = hv$. The energy E_{vib} can also be expressed be related to Boltzmann's constant K by the equation $E_{vib} = \mathrm{KT}$

then

$$hv = KT \tag{1.17}$$

or

$$v = \frac{\mathrm{KT}}{\mathrm{h}} \tag{1.18}$$

If we substitute this value for v in Eq. (1.16) we obtain

$$\mathrm{r} = \frac{K_{\mathrm{d}}\mathrm{KT}\,[\mathrm{A}]}{\mathrm{h}} \tag{1.19}$$

Because the Boltzmann's constant K in Eq. (1.19) is given by $\mathrm{R/N_A}$, Eq. (1.16) becomes

$$\mathrm{r} = \frac{K_{\mathrm{d}}\mathrm{RT}\,[\mathrm{A}]}{\mathrm{hN_A}} \tag{1.20}$$

where $\mathrm{N_A}$ is the Avogadro's number. But from the differential rate law the rate of formation of B* is given by

$$\mathrm{r} = k[\mathrm{A}] \tag{1.21}$$

where k is the rate constant.

$$\therefore \quad \mathrm{r} = \frac{K_{\mathrm{d}}\mathrm{RT}\,[\mathrm{A}]}{\mathrm{hN_A}} = k[\mathrm{A}] \tag{1.22}$$

$$\Rightarrow \quad k = \frac{K_{\mathrm{d}}\mathrm{RT}}{\mathrm{hN_A}} \tag{1.23}$$

This equation relates the rate constant for defect formation with the equilibrium constant. K_d can be obtained from thermodynamic properties of defect reactions such as the standard free energy change $\Delta\mathrm{G^o}$, the standard enthalpy change $\Delta\mathrm{H^o}$ and the standard entropy change $\Delta\mathrm{S^o}$.

We know that

$$\Delta\mathrm{G^o} = \Delta\mathrm{H^o} - \mathrm{T}\Delta\mathrm{S^o} \tag{1.24}$$

Also

$$\Delta\mathrm{G^o} = -\mathrm{RTlnK}_d \tag{1.25}$$

Therefore

$$\Delta\mathrm{H^o} - \mathrm{T}\Delta\mathrm{S^o} = -\mathrm{RTlnK}_d \tag{1.26}$$

Table 1.1 Some spectral regions of electromagnetic spectrum and effect on substances

Spectral region	Wavelength range (nm)	Effects
Microwaves	10^8–10^9	Molecular rotation
Far IR	$15–35 \times 10^3$	Molecular vibration
Near IR	780–2500	Electronic excitation
Visible	380–780	Valence electron excitation
UV	200–380	Valence electron excitation
Vacuum UV	40–200	Valence electron excitation
X-ray	0.01–10	Core-level electron excitation
γ-ray	< 10–4	Nuclear excitation

Finally we obtain

$$\ln K_d = \frac{\Delta S^*}{R} - \frac{\Delta H^*}{RT} \tag{1.27}$$

$$\Rightarrow K_d = e^{\frac{\Delta S^*}{R}}.e^{\frac{-\Delta H^*}{RT}} \tag{1.28}$$

Thermodynamic functions for defect reactions (such as ΔH^o) can be extrapolated from the values of equilibrium constant using the thermodynamic relationship in Eq. (1.29). The equilibrium constant can be measured by conventional methods which involve the measurement of electrical conductivity, thermoelectric power, and thermogravimetry (Nowotny et al. 2008).

$$\ln K_d = \frac{\Delta S^o}{R} - \frac{\Delta H^o}{RT} \tag{1.29}$$

Where K_d is the equilibrium constant for the defect reaction at equilibrium, ΔS^o is the standard entropy change during the defect formation and ΔH^o is the corresponding standard enthalpy change.

1.3.4 Photoactivation of Semiconductors

Light is a stream of oscillating electric and magnetic fields in a form usually referred to as electromagnetic radiation. In eighteenth century, T. Young and A.J. Fresnelconfirmed from separate experiments on diffraction and interference respectively, that light as electromagnetic radiation, exhibits wave character. Electromagnetic radiations have wavelengths which span a number of regions known collectively as the electromagnetic spectrum. Table 1.1 shows some spectral regions of electromagnetic spectrum and their effects on atoms or molecules. It should be clear that we were selective in choosing border wavelengths for some spectral regions as differences exist across different literature.

The regions of electromagnetic spectrum of much concern to photochemists are the UV, visible and part of very near IR (particularly 200–1,100 nm). Ultraviolet light region can be subdivided into three: UV-A covers 320–380 nm, UV-B ranges from 280 and 320 nm and UV-C refers to light having wavelength below 280 nm. The visible region is composed of sub-regions of varied wavelength ranges as red (700–620 nm), orange (620–580 nm), yellow (580–560 nm), green (560–490 nm), blue (490–430 nm) and violet (430–380 nm). Ultraviolet continuum sources include xenon arc lamp (which emits radiation of wavelength 200–1,000 nm) and H_2 and D_2 lamp (emitting 160–380 nm) while visible light continuum sources include halogen lamp and tungsten lamp (which emit radiations within 320–2,400 nm). The near IR region is not popularly used in photocatalysis although huge potentials have been established. For example, certain doped TiO_2 materials can have extended absorption edge up to 600 nm and proven efficiency for photocatalytic applications.

The absorption of light by solid was first demonstrated in 1859 by Gustav Kirchhoff (1824–1887). Gustav showed that the dark line in solar spectrum was due to the absorption of solar radiation by sodium atoms at 589 nm. In 1817, J.D.T. von Grotthus stated the first law of photochemistry which was independently proposed by J.W. Draper. The Grotthus-Draper law stresses that only light that is absorbed can be effective in producing chemical change. We can interpret this law to mean that photochemical change is not produced by mere illumination intensity but rather by the light absorbed. At molecular scale, energy is absorbed or emitted in discrete quantities called quanta which are equal in magnitude to hv (as postulated by the quantum theory). The Stark-Einstien's law of photochemical equivalence (1805) states that each molecule that takes part in photochemical reaction absorbs one quantum of light. Despite draw backs on the basis of energy loss or excessive energy absorption by reacting molecules, the law is still relevant because in general, at least a minimum of a single quantum of energy is required to cause the activation of a photocatalyst.

Upon photoabsorption of the light having sufficient frequency v, an electron is excited excitation from the occupied valence band of the semiconductor photocatalyst to the empty conduction band which results in the generation of conduction band electron (e_{CB^-}) and valence band hole (h^+_{VB}) as shown by Eq. (1.30).

$$\mathrm{SC} \xrightarrow{hv \geq E_g} \mathrm{SC}^*(\mathrm{h}^+_{\mathrm{VB}} + e^-_{\mathrm{CB}}) \tag{1.30}$$

The charge carriers produced as a result of band gap excitation have the potential to promote various photochemical events on the semiconductor surface. Figure 1.6 gives a brief schematic presentation of the processes that take place subsequent to semiconductor excitation.

Basically, photogenerated charge carriers will migrate fast within the semiconductor to the surface, whereupon they would undergo photochemical and photophysical events notably trapping and recombination. Since the migration of the carriers is fast, most of their recombination (a negative process) occurs at the surface. However, the recombination of valence band hole and conduction band electron (i.e. band-band recombination) cannot be ruled out. In Fig. 1.6, surface reactions have been enclosed in square brackets and labelled separately to avoid misconception with lattice processes. Semiconductor surfaces have the ability to lead both oxidation and reduction reactions owing to the presence of charge carriers. The reductant adsorbate or donor (D_{ads}) undergoes hole oxidation while the oxidant or acceptor (A_{ads})

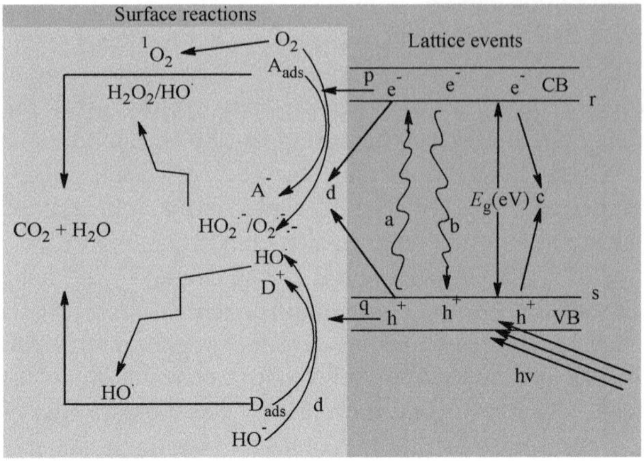

Fig. 1.6 Schematic diagram showing the central events in semiconductor photocatalytic oxidation (**a**) band gap excitation of electron, (**b**) band-band recombination, (**c**) recombination of electron and hole in the bulk, (**d**) surface recombination of electron-hole pair, (*p*) electron transfer, (*q*) positive hole oxidation, (*r*) lowest CB energy level, (*s*) highest VB energy level

undergoes electroreduction as shown by Eq. (1.31) and (1.32). The feasibility of occurrence of specific reactions can be predicted based on standard hydrogen electrode (SHE) potential and band positions of the semiconductor photocatalyst.

$$e^-_{CB} + A_{ads} \rightarrow A^-/A^{-\bullet}_{ads} \tag{1.31}$$

$$h^+_{VB} + D_{ads} \rightarrow D^+/A^{+\bullet}_{ads} \tag{1.32}$$

Electrochemically, for the reduction of any adsorbed oxidant A_{ads} to take place the conduction band must be more negative than the corresponding redox potential while for the oxidation of adsorbed reductant D_{ads} the valence band must be more positive than the redox potential [48]. These interfacial processes form the basis of photocatalytic reduction/oxidation on semiconductor particles. Redox potential plays many roles in photocatalysis. For example, the electron donor/acceptor pair in TiO_2 ought to have more than enough energy (3.2 eV $\approx$ 309 KJ mol^{-1}) to decompose water into hydrogen and oxygen, which ordinarily requires 1.23 eV, but the liberation of H_2 and O_2 is hindered by very high overpotentials (Swaddle 1990).

1.3.5 Band Gap Energies

Band gap energy corresponds to the energy difference between the lowest conduction band and highest valence band positions of the semiconductor. Band gap positions may shift with change in pH or solvent. Exact knowledge of band gap energy is very fundamental to photocatalysis to envisage the wavelength of incident light that can excite the semiconductor photocatalyst (Swaddle 1990). The band gap energies of some common semiconductor photocatalysts at pH 0 are shown in Fig. 1.7. Experimental band gap energies may vary from the standard band gap energies in the figure depending upon the method preparation of the semiconductor.

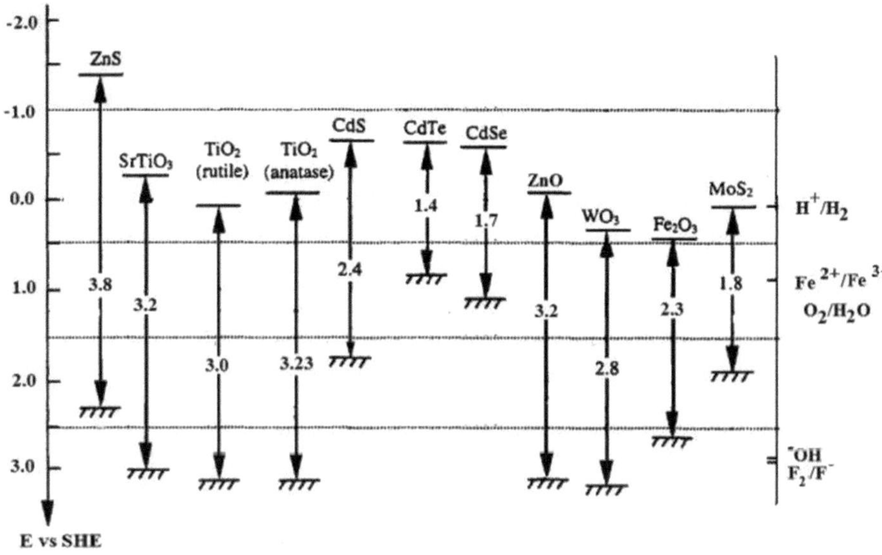

Fig. 1.7 Band gap energies (eV) of representative semiconductor photocatalysts against normal hydrogen scale at pH 0

The band gap of semiconductor photocatalyst powders has been determined variously by diffuse reflectance spectroscopy (DRS). Primarily, the reflectance (R) of specimens is measured against a reference diffuse reflector (such as spectralon and $BaSO_4$) using UV/Vis/NIR diffuse reflectance spectrophotometer. The band gap (E_g, eV units) can be calculated from the threshold/cut-off absorption wavelength (λ, nm) of the sample by the energy-wavelength relationship (Feng et al. 2009). The energy of a photon of frequency v and wavelength λ may be given by:

$$E = hv \quad (1.33)$$

$$E = \frac{hc}{\lambda} \quad (1.34)$$

where v is the light frequency. If the values of h (Plank's constant = 6.626×10^{-34} Js) and c (speed of light = $2.988 \times 10^{8}\text{ms}^{-1}$) are plugged into the equation we obtain a simplified expression for the band gap energy:

$$E_g(\text{J}) = \frac{19.8\text{x } 10^{-17}}{\lambda} \quad (1.35)$$

Using the above equation, the energy absorbable at any given light wavelength can be estimated. If the band gap energy is desired to be calculated in eV units, where 1 eV = 1.6×10^{-19} J, the following equation may be used to calculate band gap:

$$E_g(\text{eV}) = \frac{1237.5}{\lambda} \quad (1.36)$$

It may be appropriate to define the parameter E' which refers to energy (in J mol^{-1}) per einstien (6.022×10^{23} photons) had often been used in photochemical studies. This parameter is given by Eq. (1.37).

$$E' = \frac{N_A hc}{\lambda} \tag{1.37}$$

Where N_A is the Avogadro's number of photons.

Actually, the optical band gap (E_g) of a semiconductor solid can be more accurately determined from the intercept of the extrapolated tangent of the plot of logarithmic remission function ($\ln\alpha/S$) against photon energy absorbed (hv), where α and S represent absorption and scattering coefficient, respectively. The ratio α/S can be calculated from the experimentally measured reflectance by Kubelka-Munk equation (Trikalitis et al. 2001):

$$\frac{\alpha}{S} = \frac{(1-R)^2}{2R} \tag{1.38}$$

The ratio α/S is referred to as Kubelka-Munk or remission function. Since the Kubelka-Munk ratio is a function of reflectance, it is interchangeably used with F(R). Assuming remarkably effective absorption then $\alpha = \mathrm{F(R)}$. The absorption coefficient α varies with the light energy absorbed according to Tauc's equation (Tauc and Menth 1972; Kako et al. 2008):

$$\alpha hv \propto (hv - E_g)^n \tag{1.39}$$

So,

$$(\alpha hv)^{1/n} \propto (hv - E_g) \tag{1.40}$$

where n is a constant accounting for the type of optical transition.

Equivalently,

$$(\mathrm{F(R)}hv)^{1/n} \propto (hv - E_g) \tag{1.41}$$

The Tauc band gap can be obtained as intercept of the plot of $(\alpha hv)^{1/n}$ or $(\mathrm{F(R)}hv)^{1/n}$ against hv. The values of n are 1/2, 2, 2/3, 3 for allowed direct, allowed indirect, forbidden direct, and forbidden indirect transitions, respectively. For instance if we assume direct optical transition for ZnO and that remarkably effective absorption occurs then $n = 1/2$ and $\alpha = \mathrm{F(R)}$, respectively. We have recently obtained the band gap of ZnO powder from Merck to be 3.36 (Fig. 1.8) (Gaya 2011). This value was checked by the plot of $(\alpha hv)^2$ or $(\mathrm{F(R)}hv)^2$ against hv where it was obtained as intercept (Fig. 1.8, inset).

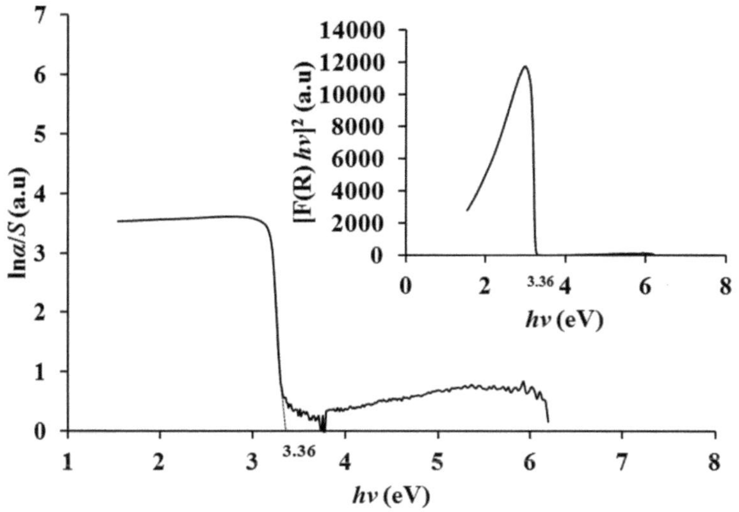

Fig. 1.8 Kubelka-Munk spectral function of ZnO photocatalyst versus excitation energy (Reproduced from Gaya 2011). The inset is a Tauc plot depicting the variation of $(F(R)h\nu)^2$ against light photons absorbed. Diffuse reflectance measurements were performed on Perkin Elmer Lamda 35 UV/Vis/NIR spectrometer

1.4 Photocatalytic Processes

In an attempt to classify photocatalysis, we can roughly distinguish two broad types: photocatalytic synthesis and photocatalytic degradation. This classification is informed by the fact that photocatalytic reactions can be directed either towards formation of new substances or decomposition of reacting material. The term photocatalytic degradation will encompass photosplitting of water, photodisinfection and photodetoxification of inorganic materials whereas photocatalytic synthesis will cover the photocatalytic reduction of CO_2 to organic compounds, organic synthesis and photocatalytic fixation. Generally, the primary reactions that may take place en route photocatalytic reactions depend upon the chemical nature of the substrate and the reaction condition. Basically for the purpose of effective photocatalytic synthesis, oxidations are carried out in presence of partially hydrated (Fox et al. 1989) or neat solvent (Ohuchi et al. 2007). A vast majority of the photosynthesis reactions that take place on the surface of semiconductor photocatalyst are characterised by positive ΔG^o which means therefore they are thermodynamically favourable. In fact, the conversion of several functional group has been successfully studied by this approach since 1980s (Fox and Chen 1983) and often this is associated with high selectivity (Fox et al. 1989).

On the other hand, the aqueous phase transformation of substrates eventually results in total decomposition of the organic material hence befits environmental remediation application. This type of photocatalysis characterised by negative ΔG^o.

In both photocatalytic degradation and synthesis, dioxygen is invariably required for oxidation (Cameron and Bocarsly 1985). However, photocatalytic synthesis by reduction processes can proceed in absence of oxygen. Both the reductive and oxidative approaches to organic oxidation offer useful mechanistic information. In the aqueous phase oxidative transformations, the mediators are several in-situ generated oxidising species, the most effective among them being hole and hydroxyl radical. These oxidising species involve organic adsorbates in oxidation by direct or indirect hole transfer pathways. The indirect pathways involve reactive radical species to cause bond fission. Generally, the mechanism for photocatalytic reactions may comprise several intermediary steps such as hydrogen abstraction, hydroxylation and bond scission resulting in functional group conversion. Some of these steps are shown below:

Decomposition:

Photocatalytic hydrogen production is based on the decomposition of water on suitable semiconductor photocatalyst:

$$2H_2O \rightarrow 2H_2 + \frac{1}{2}O_2 \tag{1.42}$$

Hydroxylation:

Ring-based reactants are easily attacked by hydroxyl radicals. In case ring opening did not result, hydroxylation of the ring occurs. For example, phenolic compounds get hydroxylated upon hydroxyl radical attack system as given in the following equation:

$$\text{Ph-OH} + HO^{\bullet} \rightarrow (HO)_2\text{Ph}{+}H_2O \tag{1.43}$$

Oxygen addition:

The possibility of peroxidation of radicals in photocatalytic solutions has also been reported. In such a case, a hydroxyphenyl radical is oxidised to hydroxyphenyl peroxyl radical.

$$\text{HOPh}^{\bullet} + O_2 \rightarrow \text{HOPhO}_2^{\bullet} \tag{1.44}$$

Dimerisation:

The abstraction of hydrogen is also common in photocatalysis. The formation of alkyl and hydroxyphenyl radical is shown in the chemical equations below.

$$\text{R-H}{+}HO^{\bullet} \rightarrow R^{\bullet} + H_2O \tag{1.45}$$

$$\text{Ph-OH}{+}HO^{\bullet} \rightarrow \text{HO-Ph}^{\bullet} + H_2O \tag{1.46}$$

Dimerisation:

The dimerisation of alkyl or phenyl radical has been an established step in the mechanism of photocatalytic reactions. In the process, radicals undergo termination reaction to yield higher molecular hydroxybiphenyl.

$$\text{HO-Ph}^{\bullet}{+}HO^{\bullet} \rightarrow \text{HOPh-PhOH} \tag{1.47}$$

Functional group conversion:

Some many functional group conversions can be facilely achieved by semiconductor-assisted photocatalysis. These include the reduction of dichlorophenoxyacetic acid to dichlorophenol:

$$HOOCCH_2O\text{-}Ph(Cl)_2 \rightarrow HO\text{-}Ph(Cl)_2 \quad (1.48)$$

It is a common knowledge that many nitrogen oxides in the atmosphere are formed via photocatalytic conversions of N_2. In particular, photocatalytic nitrogen fixation and the reverse process are known to proceed photocatalytically:

$$N_2 \rightleftharpoons NH_3 \quad (1.49)$$

Other forms of reaction induced in presence of semiconductor photocatalysts include nitrification (Eq. (1.50)):

$$NH_3 \rightarrow NO_3^- \quad (1.50)$$

In a more clearly-defined and systematic way, photocatalytic reactions can be distinguished based on redox feature into photocatalytic oxidation or reduction. This classification would be more suitable for describing all forms of reactions under photocatalysis. In environmental context, the use of various semiconductor photocatalysts for the photoreduction of several metal ions has been well established despite the fact that the photoreduction of metal ions is pH dependent. Considerable reactions rates have been obtained in the environmental ranges of concentration and pH. Remarkable photoreduction (99.3 % in 1 h) of Cr^{5+} (in concentrations of 200 $\mu g ml^{-1}$) was obtained in presence of CdS photocatalysts (Wang et al. 1992). The photocatalytic reduction of Hg^{2+} to Hg^0 over different photocatalysts has also been reported (Wang et al. 2004). Within 30 min, about 100 mg dm^{-3} Hg^{2+} can be removed from solutions containing TiO_2 at pH 4. Composite of ZnO-WO_3 photocatalyst has demonstrated similar activity for the removal of Hg^{2+} (Wang and Zhuang 1993).

Enhancement of either photocatalytic oxidation or reduction can be achieved positively in the presence of suitable organic molecules added as sacrificial agents. Col∉n et al. (2001) observed higher photoreduction rate of Cr^{6+} over TiO_2 in presence of non-aromatic carboxylic acids. With aromatic carboxylic acids however, catalyst deactivation was observed. It may be imperative during such studies to take into account the possibility of interaction of the metal ion with the pollutant as Elovitz and Fish (1994) have confirmed the interaction of fourteen phenolic compounds with Cr^{6+} which led to the reduction of the latter to Cr^{3+}.

1.5 Chemistry of the Inception Reactions

Immediately produced, charge carriers migrate through the bulk semiconductor particle. These delocalised charge carriers can be trapped by physisorbed water molecules, surface oxygen vacancies (or generally defects) and surface hydroxyl groups (Serpone 1995; Zhang et al. 2009). In addition, the charge carriers get annihilated as they come close to one another by a process known as recombination. Photocatalysis thrives only when recombination processes are reduced to the barest minimum. Basically, the average transit time for the trapping of charge carrier τ_{tr}, within a particle with radius r can be obtained by solving Fick's of diffusion law.

$$\tau_{tr} = \frac{r^2}{\pi^2 D} \tag{1.51}$$

where D is the diffusion coefficient. Typical values of $D = 0.1\ \text{cm}^2\text{s}^{-1}$ and $r = 10\ \text{nm}$ which gives average transit time of about 1 ps, much shorter than the recombination time, so that most charge carriers can reach the surface before recombination (Robertson et al. 2005). Unlike recombination which is a physical process, trapping takes place by a chemical mechanism which involves charge carrier transfer processes.

1.5.1 Charge Carrier Trapping

Sufficient evidence for trapping of electrons may be traced back to the work of Ward and Bard (1982) in which trapping of electrons was observed in TiO_2 suspensions containing platinum deposits. Later on Anpo and Kubokawa (1984) demonstrated the trapping of charged carriers in metal and oxygen site of the semiconductor oxide and their role in photoinduced processes. In a parallel investigation, Bahnemann et al. (1984) provided an evidence for sub-surface trapped holes using TiO_2 sol. Chemically, the trapping of charge carriers may be represented by the reactions in Eq. (1.52)–(1.55).

Trapping of e^-_{CB} *at defects*:

$$\text{Ti}^{+4}\text{OH} + e^-_{CB} \rightarrow \text{Ti}^{+3}\text{OH} \tag{1.52}$$

Trapping of h^+_{VB} *at defects*:

$$\text{Ti}^{+3}\text{OH} + h^+_{VB} \rightarrow \text{Ti}^{+4}\text{OH} \tag{1.53}$$

Trapping of h^+_{VB} *at defects*:

$$\text{Zn}^{+2}\text{O} + h^+_{VB} \rightarrow \text{Zn}^{+2}\text{O}^{\bullet} \tag{1.54}$$

Surficial trapping of e^-_{CB}:

$$-\text{OH}^-_{surf} + e^-_{CB} \rightarrow \text{OH}^{\bullet}_{ads} \tag{1.55}$$

The nature of trapped electrons was not fully comprehended. However, Energetically-based description involving trap sites has recently been relied upon, thanks to the work of Bahnemann et al. (1984). Thus, an electron or hole may either be deeply

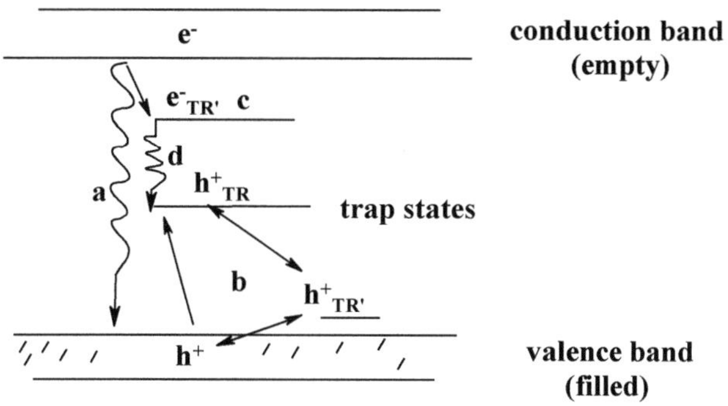

Fig. 1.9 Conceptual diagram showing (**a**) band-band recombination (**b**) hole trapping triangle (**c**) shallowly trapped electron (**d**) trapped-assisted recombination

entrapped in energy levels as $e^{-}{}_{TR}/h^{+}{}_{TR}$ or shallowly trapped as $e^{-}{}_{TR'}/h^{+}{}_{TR'}$ in the semiconductor states as shown by Eq. (1.56) and (1.57).

$$e^{-}_{CB} \rightarrow e^{-}_{TR}/e^{-}_{TR'} \quad (1.56)$$

$$h^{+}_{VB} \rightarrow h^{+}_{TR}/h^{+}_{TR'} \quad (1.57)$$

Electrons or holes may undergo several activities while trapped in the bulk of the semiconductor and as they approach the surface. Trapping levels exist which may be considered as those levels corresponding to conduction band energies or valence band energies. Electrons are likely to be trapped at the centre of the energy gap or near conduction band known as electron traps whereas positive hole may be captured at hole traps near the valence band. Figure 1.9 shows the trapping of charge carriers and their recombination in the bulk. Both the shallow and deeply trapped states are shown in the figure. In addition, we used a triangular path to highlight the movement of the hole into shallow trap, then deep trap or directly to a deep trap.

On the other hand, it has unanimously been accepted that deeply-trapped hole is chemically the same as surface-bound hydroxyl radicals (Bahnemann et al. 2003). Since these hydroxyl radicals are readily reactive the possibility of diffusion-controlled kinetics may be ruled out in any efficient photocatalytic system.

1.5.2 Primary Electron and Hole Transfer Reactions

Our understanding of photochemistry of semiconductor-based photo-oxidation is strongly rooted in the concept of the formation of oxidising species and the resulting surface reactions. In addition to positive hole, several reactive oxygen species such as such as $O_2^{\bullet -}, OH^{\bullet}, HO_2^{\bullet}$ and H_2O_2, are reported to be strong oxidising agents in photocatalytic solutions. Unlike in photocatalytic synthesis (Ohuchi et al. 2007), the presence of water as a source of hydroxyl radical is an essential initial step that leads to photocatalytic degradation. Either water molecule, terminal oxygen ($> O^{2-}_{term}$) or terminal hydroxyl group ($> HO^{-}_{term}$) may be oxidised by positive hole to form hydroxyl radical (Eq. (1.58)–(1.61)) (Imoberdorf et al. 2005; Haick and Paz 2003) respectively. In this regard, a minor view is worth noting which stresses based on electron photoemission spectroscopy techniques data reported in the literature that surface adsorbed water cannot be oxidised by band gap UV light (Salvador 2007). Whatever is exactly the mechanism, water is the source of hydroxyl radical and the initial oxidant is the positive hole.

$$H_2O_{ads} + h^+_{VB} \rightarrow HO^{\bullet}_{ads} + H^+ \quad (1.58)$$

$$H_2O_{(aq)} \xrightarrow{autoionisation} HO^-_{(aq)} + H^+_{(aq)} \quad (1.59)$$

$$HO^- + h^+_{VB} \rightarrow HO^{\bullet} \quad (1.60)$$

$$> O^{2-}_{term} + H^+_{(aq)} + h^+_{VB} \rightarrow > HO^{\bullet}_{term} \quad (1.61)$$

The generation of hydroxyl radicals in photocatalytic processes has been confirmed by electron paramagnetic resonance (EPR) technique (Peter et al. 1997). It was found that at low intensities the rate of formation of $HO^{\bullet}$ was a linear function of light intensity. Even though dioxygen can influence the rate of photocatalytic oxidation it has less effect on the generation of hydroxyl radicals. This fact supports the photohole pathway for the generation of hydroxyl radical. It should be noted however that the electron pathway has also been reported and this will be discussed infra. Even though adsorbed water cannot be excited by UV light (Salvador 2007), vacuum UV-photolysed water existing on the surface of TiO_2 may likely decompose homolytically to yield hydroxyl radical at 185 nm wavelength (Yoon et al. 2008).

There have been several attempts to quantify the yield of reacting oxygen species. The yield of adsorbed hydroxyl radical ($HO^{\bullet}_{ads}$) in irradiated aqueous TiO_2 suspension has been determined to be nearly constant across a wide pH range (Du and Rabani 2003; Gao et al. 2002). Therefore, pH can only affect the adsorption equilibrium of the ionic substrates but not the generation of hydroxyl radicals . Since it is native, adsorbed hydroxyl radical is more likely to remain in the adsorbed phase. However, there is possibility of diffusion of a significant amount of adsorbed hydroxyl radicals into the bulk solution which will react as free hydroxyl radicals ($HO^{\bullet}_{free}$) especially with surface-sensitised photocatalysts (Minero et al. 2000).

Hydrogen peroxide is constantly produced during the irradiation of semiconductor oxide dispersions. When water dissociates on the surface of semiconductor particle (Eq. (1.52)) OH^- is adsorbed and the H^+ is consumed in the process of hydrogen peroxide formation. The formation of hydrogen peroxide proceeds by a reduction mechanism which involves the generation of $HO^{\bullet}_2$ (De Lasa et al. 2005). Since water undergoes poor ionisation, only limited H_2O_2 may be formed by this process. In the initial stage of the process, adsorbed molecular oxygen is reduced to superoxide radical by reaction with trapped electrons at the defects of the photocatalyst (Eq. (1.56)) which is characterised by high rate constant (7.6×10^7 dm^3mol^{-1} s^{-1}) (Bahnemann et al. 1997).

In photodegradation reactions, the norm is to continuously supply oxygen for efficient destruction of persistent organic substances. The oxygen will adsorb on the semiconductor photocatalyst. This adsorbed oxygen has the potential to initiate photoreaction (Lu et al. 1995) by the formation of superoxide radical (Eq. (1.62)).

$$O_{2ads} + e^-_{CB} \rightarrow O^-_{2ads} \tag{1.62}$$

The resulting superoxide radical reacts as Brønsted base to form hydroperoxyl radical as shown in Eq. (1.66). This process is very important as it would circumvent the electrons necessary for recombination and eventually lead to the formation of hydrogen peroxide. Efficient utilisation of electrons such a process can be realised with the aid of promoters such as Pt (Ranjit and Viswanathan 1997; Sun et al. 2005).

$$O^{-\bullet}_{2ads} + H^+ \rightarrow HO^{-\bullet}_{2ads} \tag{1.63}$$

The hydroperoxyl radical is believed to undergo the following elementary processes to result in the formation of hydrogen peroxide:

$$HO^{-\bullet}_{2ads} + e^-_{CB} \rightarrow HO^{-\bullet}_{2ads} \tag{1.64}$$

$$HO^{-\bullet}_{2ads} + H^+_{ads} \rightarrow H_2O_{2ads} \tag{1.65}$$

Besides the reductive pathway, the formation of hydrogen peroxide may also involve an oxidative pathway based on ${h^+}_{vb}$. However, the dominance of the reduction mechanism (Eq. (1.64) and (1.65)) has been confirmed (Horváth 2003). The population of conduction band electrons for the overall process can be maximised by the addition of hole scavenger. For instance, it has been demonstrated in presence of carboxylic acids, ZnO and suitable hole scavengers such a reduction process can be enhanced up to 2×10^{-3} mol dm^{-3} H_2O_2 (Hoffman et al. 1994). The rate of H_2O_2 formation in photocatalytic processes can be monitored based on quantum yield for H_2O_2 production and quantum yield for H_2O_2 destruction. Hoffmann et al. (1995) have provided the relationship below for the calculation of the rate of H_2O_2 formation:

$$\frac{d[H_2O_2]}{dt} = (\Phi_o\text{-}\Phi_1[H_2O_2])\left(\frac{d(h\nu)}{dt}\right) \tag{1.66}$$

Where d[H_2O_2]/dt is the rate of H_2O_2 formation and Φ_o is the quantum yield for H_2O_2 production and Φ_1 is quantum yield for H_2O_2 destruction.

1.6 Roles of Reactive Oxidising Species

1.6.1 Influence of Hole Juxtaposed to Hydroxyl Radical

While the benefits of the generation of hole is crucial to photocatalytic degradation, it still remains unclear whether the positive hole or hydroxyl radical is the main oxidant, and if there is any dependence on the kind of compound being investigated. Basically,

Table 1.2 Oxidation potential of fluorine and photooxidants with typical rate constants in advanced oxidation of organics

Oxidising agent	E^o (V)	Typical rate constants ($mol\ dm^{-3}s^{-1}$)
$HOO^•$	+1.70	–
ClO_2	+1.57	–
H_2O_2	+1.78	
O_3	+2.07	10^{-2}–10^3
$HO^•$	+2.80	10^6–10^{11}
F_2	+2.87	–

it is widely believed that hydroxylated ring products are formed from the reactions of aromatic compounds mainly via the hydroxyl radical mechanism. In support of this opinion, hydroxyl radical has been shown experimentally to be the major oxidant involved in 70–90 % of the photo-oxidation of cyclic organic compounds notably chlorophenols (Richard and Boule 1991; Sehili et al. 1991). The photocatalytic transformation of phenol on TiO_2 and TiO_2/F (0.01 mol dm^{-3} F^-) at pH 3.6 in the presence of different alcohols (*tert*-butyl alcohol, 2-propanol, and furfuryl alcohol) proceeded through the reaction with surface bound hydroxyl radical by 90 %, the remaining 10 % via a direct hole attack (Minero et al. 2000).

Of course, hydroxyl radical has higher oxidising power than all of the oxidising species exploited in advanced oxidation processes such as hydroperoxyl/superoxide radical, hydrogen peroxide and ozone, which may be evident from its high oxidation potential. Only fluorine supersedes hydroxyl radical in terms of oxidation potential which is one of the reasons why many environmental pollutants can be abated by this radical. Table 1.2 shows the standard oxidation potential (E^o) of some strong chemical oxidants encountered in photocatalysis (Serpone and Pelizzetti 1989; Patnaik 2004; Housecroft and Sharfe 2005; Litter 2005; Shiying et al. 2009; Blanco et al. 2009). It would be seen from the table that the typical rate constant for advanced oxidation by hydroxyl radical is folds of that involving ozone.

There are four ways by which hydroxyl radical may cause oxidation of organic pollutant (Turchi and Ollis 1990). Adsorbed hydroxyl radical can oxidise nearby pollutant molecules. Conversely, free hydroxyl radicals can oxidise adsorbed organic molecule. Both of these processes may be classed under the Langmuir-Rideal bimolecular mechanism. Thirdly, oxidation can take place while both the hydroxyl radical and the pollutant are in the adsorbed phase in accordance with the Langmuir-Hinshelwood mechanism. Finally, free hydroxyl radicals may react with pollutant molecules in solution. It is important to note that, the last process does not usually occur in photocatalytic systems as several workers have given substantial evidence for the occurrence of photocatalytic reactions solely at the adsorbed phase. The remaining three mechanisms can be pictorially represented by the scheme in Fig. 1.10.

A different account of oxidation mechanisms by Draper and Fox provided an ultimate experimental evidence to support the direct hole oxidation mechanism in photocatalysis (Draper and Fox 1990). These workers could not detect any of the expected intermediate of hydroxyl radical pathway by diffuse reflectance and transmission flash photolysis study. A priori, it is commonly known that organic pollutants

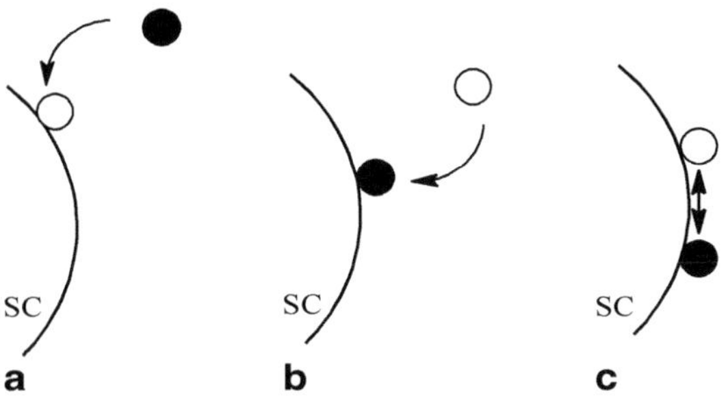

Fig. 1.10 Bimolecular schemes for the reaction of pollutant molecule with hydroxyl radical in presence of semiconductor particle SC. (**a**) pollutant in the vicinity of semiconductor surface (SC) attacked by adsorbed hydroxyl radical (**b**) oxidation of pollutant while at the surface of semiconductor photocatalyst and the hydroxyl radical in solution (**c**) oxidation reaction while both hydroxyl radical and pollutant are in the adsorbed phase. • target molecule, ○ hydroxyl radical

as well as the hydroxyl ion can be oxidised directly by hole. Moreover, quantum yield study by Ishibashi et al. (2000) concluded that the oxidative mechanism over semiconductor ought to predominantly proceed via positive hole mechanism. It may be difficult to outrightly reconcile the foregoing discrepancies. We however recommend further investigations over a wide range of classes of compounds.

Chemical compounds under photocatalytic degradation do not yield the same way. Moreover, both hole and hydroxyl radical are competing oxidants in photocatalytic processes. In fact, some evidence has been found for the involvement of both positive hole and hydroxyl radical in photocatalytic oxidation (Richard 1993; Stafford et al. 1994; Sclafani and Herrmann 1998; Balcerski et al. 2008) Therefore, the dominant pathway during photocatalytic degradation may depend upon the nature of the substrate. For example, the degradation of phenolics above has been explained largely on the basis of hydroxyl radical mechanism. Similarly, the role of primary active species in the degradation of acetaminophen (i.e. paracetamol) over TiO_2 was investigated by Yang et al. (2009) and hydroxyl radical was identified to be the main oxidant. On the whole, it may be rational to say the intensification of research on the role of hydroxyl radical or hole with different classes of compounds for a wide range of compounds merits much attention in order to provide specific measure of structure-performance relationship.

1.6.2 Superoxide Radical

Superoxide radical has been suggested to play dominant role in photo-oxidation of certain pollutants particularly under visible-light irradiation (Wang et al. 2009). Fu et al. (2006) gave the first direct evidence that $HO^{\bullet}$ and $O_2^{-\bullet}$ are responsible for the decomposition of 4-chlorophenol over N-doped TiO_2 under visible light illumination, using electron spin resonance (ESR) spin trapping technique. These workers also observed low production of $HO^{\bullet}$and$O_2^{-\bullet}/HO_2^{\bullet}$ in N-doped TiO_2 compared to bare TiO_2.

Similarly, it has been shown that the conversion of NH_3 on TiO_2 surface involves the oxidation of $NH_2^{\bullet}$ by superoxide radical to NO^- which is further oxidised to NO_x^- ($x = 2$ or 3) by reaction with oxygen (Eq. (1.67)–(1.69)) (Yamazoe et al. 2007).

$$O_{2ads}^{-\bullet} + NH_2^{\bullet} \rightarrow NO^- + H_2O \tag{1.67}$$

$$2NO^- + O_2 \rightarrow 2NO_2^- \tag{1.68}$$

$$NO^- + O_2 \rightarrow NO_3^- \tag{1.69}$$

Besides direct involvement in photooxidation, the efficient formation of superoxides would mean reduced recombination rate as this oxidising species is formed via electron transfer to oxygen molecule. In fact, the generation of superoxide is a process considered very important as it can control the photocatalytic reaction rate.

1.6.3 Hydrogen Peroxide

Hydrogen peroxide can affect photocatalytic reactions favourably or unfavourably depending upon its levels in solution vis-à-vis substrate and on the semiconductor photocatalyst. For instance, hydrogen peroxide generated in-situ in titania dispersions showed favourable effect on the rate of degradation of 1,2-dimethoxybenzene depending on $[H_2O_2]/[1,2\text{-dimethoxybenzene}]$ ratio (Amalric et al. 1995). With ZnO however, negative results were obtained. The mechanism for rate enhancement involves the reduction of hydrogen peroxide to hydroxyl radical either through direct electron transfer (Eq. (1.73)) or by Haber-Weiss reaction (Eq. (1.74)) (Amalric et al. 1995; Hirakawa and Nosaka 2002). Hence, both hydrogen peroxide and dioxygen molecule compete for electron transfer (Auguliaro et al. 1990).

The reduction of hydrogen peroxide to hydroxyl radical and dioxygen to superoxide on TiO_2 were confirmed by luminol chemiluminescence and terephthalic acid fluorescence probe method respectively, although Haber-Weiss reaction was calculated to be negligible in hydroxyl radical generation (Hirakawa and Nosaka 2002). Within optimum levels, added hydrogen peroxide is known to increasingly enhance degradation rate of organic pollutants in TiO_2 irradiated system (Amalric et al. 1995). In the absence of TiO_2 the efficiency of H_2O_2 is reduced due to self photocatalytic degradation (Augugliaro et al. 1990). The decomposition of hydrogen peroxide was found to be proportional to the intensity of light applied (Ilisz et al. 1998).

$$H_2O_2 + e_{CB}^- \rightarrow 2HO^{\bullet} \tag{1.70}$$

$$H_2O_2 + O_2^{-\bullet} \rightarrow {}^{\bullet}OH + OH^- + O_2 \tag{1.71}$$

Unlike in ZnO and CdS irradiated systems, hydrogen peroxide is not readily detected in bare TiO_2 dispersions due to its immediate removal as it is formed (Harbour et al.

1985; Ranjit and Viswanathan 1997). Several mechanisms have been suggested to account for the removal of hydrogen peroxide from TiO_2 photoreactivity dispersions among which reaction with hydroxyl radical is widely accepted (Amalric et al. 1995; Mrowetz and Selli 2006). Each mole of hydroxyl radical withdraws one mole hydrogen peroxide from the heterogeneous solution which finally results in the formation of O_2^- (Rodríguez and Fernández-García 2007). These reactions are shown by Eq. (1.72) and (1.73).

$$HO^{\bullet} + H_2O_2 \rightarrow HO_2^{\bullet} + H_2O \quad (1.72)$$

$$HO_2^{\bullet} \rightarrow H^+ + O_2^- \quad (1.73)$$

Similarly, it has been demonstrated that an overdose of hydrogen peroxide on TiO_2 particles will retard photo-oxidation rate (Chu and Wong 2004), which may also be as a result of the competition of the hydrogen peroxide and the organic molecule for hydroxyl radical. So, the above mechanism can be used to explain decrease in reaction rate with exceeding levels of H_2O_2.

1.6.4 Singlet Molecular Oxygen

Photoproduced $O_2^{-\bullet}$ is likely to be converted to lowest excited state of molecular oxygen$^1O_2(\Delta g)$. Consequently, there is possibility that the latter lends support in the photocatalytic oxidation of pollutants on semiconductor photocatalysts. Singlet oxygen ($^1O_2$) is a powerful oxidant that can even mediate the photo-oxidation of alkenes (Zhang et al. 2003). However, there has been only a few reports indicating experimentally, the formation of $^1O_2$ on TiO_2 photocatalyst.

In 1970, Khan (1970) attempted the elucidation of the nature of TiO_2 photoexcitation. He observed singlet oxygen signals at 355 nm as a consequence of TiO_2 excitation. Recently, Daimon and Nosaka (2007) detected singlet molecular oxygen from a TiO_2 photocatalyst irradiated with a 355 nm laser pulse by monitoring its near-infrared phosphorescence at 1,270 nm. The quantum yield was significant (0.12–0.38 in air) while the lifetimes ranged from 2.0 to 2.5 μs which suggests that$^1O_2$ may also contribute in the photo-oxidation of organic molecules. An examination of eighteen TiO_2 samples revealed increase in the formation of$^1O_2$ with decreasing size of TiO_2 particles, indicating that a large specific surface area causes a higher possibility of reduction producing $O_2^{-\bullet}$ $O_2^{-\bullet}$ and then a large amount of the $^1O_2$ is formed (Daimon et al. 2008).

Having established the presence of singlet oxygen in semiconductor photocatalytic systems, it is left only the study of the role of this oxidant in semiconductor photocatalysis. Surprisingly, the only study that reported the production of singlet oxygen and its role in photocatalytic oxidation was perhaps that of Jańczyk et al. (2006). These workers have given strong evidence for the degradation of cyanuric acid on F-TiO_2 by singlet oxygen, which opens new discussion on the role of oxidising species in photocatalytic oxidation.

1.7 The Fate of Inorganic and Organic Contaminants

Among photocatalytic reactions, oxidation is the most widely studied owing to its efficiency in environmental decontamination. The beauty of photocatalytic oxidation is that only mineral products form at the end of the process. This property offered photocatalytic oxidation several applications including in water and air decontamination, self-cleaning surfaces and organic carbon analysers. These variant applications are absolutely related by a common principle often referred to as photomineralisation, on the grounds thatduring the abatement of organic pollutants containing heteroatoms the corresponding benign anions are formed. Compounds containing substituents as chlorine (-Cl), bromine (-Br), fluorine (-F), iodine (-I), nitrogen (=N-, N), phosphorus (P, sulphur and cyanide form the corresponding anions as chloride (Cl^-), bromide (Br^-), fluoride (F^-), iodide (I^-), nitrate (NO_3^-), phosphate PO_4^{3-}, sulphate (SO_4^{2-}) and cyanate (OCN^-) respectively (Naman et et al. 2002; Ravichandran et al. 2007). At the other end, the abatement of hazardous metals is based on change to a less or non- hazardous oxidation state. On the whole, mineralisation speaks story about the greenness of photocatalysis as a modern treatment technology.

Naturally, the mineralisation of organic compounds consisting of many carbon atoms usually proceeds through the formation of organic intermediates. Finally CO_2, H_2O and possibly other mineral products may be formed depending upon the starting material. Overall, we may comprehensively represent the photo-oxidative mineralisation of all substituted organic compounds by Eq. (1.74).

$$\begin{aligned} C_aH_bX_mY_n + \left(2m + s + \tfrac{b-m}{2}\right) O_2 \rightarrow aCO_2 + \left(\tfrac{b-m}{2}\right) H_2O \\ + mH^+ + mX^- + mYO_s^{v-} \end{aligned} \qquad (1.74)$$

X stands for each of the heteroatoms as Cl, Br, F, Y represents I, –CN, P, S and N. *a*, *b*, *m* and *n* are the number of C, H, X and Y in the molecule. YO_s^{V-} represents any radical such as $OCN^-, PO_4^{3-}, SO_4^{2-}, NO_3^-$ having charge *v*, and *s* represents the number of O atoms in the radical. When a product is not expected, if the heteroatom is not present in the starting material for example, the term representing it on both sides of the equation can be removed. The disappearance of organic carbon in photocatalytic processes has often been confirmed by estimating the diminution in total organic carbon levels, the detection of CO_2 liberated or by monitoring the disappearance of original compound and its intermediates using chromatography. These methods have been discussed briefly in Chap. 3.

Quite generally, the importance of photocatalytic oxidation in environmental context cannot be overemphasised. Panoramic literature on concept and applications of photocatalysis along with its corollary are available and currently are on the increase. We encourage the reader to visit the references in Table 1.3 for further reading.

Table 1.3 Recommended reading materials on photocatalysis

For visible light-activated catalysts, pilot-scale industrial reactors	(Adesina 2004)
For current adsorption processes used in water purification	(Jiuhui 2008)
For overview of photocatalysis, kinetics, disinfection, regeneration of photocatalyst; metal deposition and recovery as potential application of photocatalysis	(Ollis 2000)
For Fe-TiO_2 and platinised TiO_2 photocatalyst. Novel explanation to huge increase in quantum yield such by antenna mechanism	(Wang et al. 2006)
For photocatalytic degradation of air- and water- based organics such as dyes; kinetics in comparison with other advanced processes	(Rajeshwar et al. 2008)
Preparation of semiconductor photocatalyst. Dark adsorption and influence in further reaction. Reactor configurations	
Overview of the mechanism of photocatalytic mineralisation	(Serpone and Khairutdinov 1996)
Electronic, structural and surface features of TiO_2	(Alfano et al. 2000)
For review of photocatalytic reactors with focus on utilisation artificial or solar light	
For review and engineering analysis of photocatalytic air cleaning. Report of experimental and cost studies by commercial companies	(Tompkins and Anderson 2001)
For major developments in the application of photocatalysis	(Fujishima et al. 2007)
For overview of endocrine disrupting compounds removed by photocatalysis and the operating parameters	(Belgiorno et al. 2007)
For highlights on the photocatalytic degradation of N-heterocyclic compounds, mechanism and products	(Calza et al. 2005)
For mild selective photocatalytic reactions, solar detoxification, general aspects of photocatalysis and challenges	(Braslavsky 2007)
For functional group transformation by semiconductor photocatalysis	
Mechanism of photocatalysis. Photosensitisation	
Semiconductor surface modification	(Herrmann et al. 2007)
For photocatalysis before and after Honda-Fujishima effect. Details of industrial applications of photocatalysis	(Hashimoto et al. 2005)
For general background and basic features of photocatalysis with specific examples. Update on compounds degraded	(Hoffmann et al. 1995)
For general misconceptions about quantum yield, band diagram, platinised TiO_2, reaction rate, cationic doping, Langmuir-Hinshelwood rate constants and side reactions	(Herrmann 2010a; Herrmann 2010b)
For heterogeneous photocatalysis: influence of parameters, including oxygen pressure. Mild oxidation and mineralisation	
Other topics include the treatment of industrial waste waters, noble metal recovery, solar plants	(Herrmann 2005)
For review of photocatalysis, advantages of and drawbacks. Overview of work done by the year of publication	(Matthews 1992)
For authoritative review covering major topics of TiO_2 photocatalysis and its mechanism	(Gogate and Pandit 2004)
For progress in the application of TiO_2, introducing superhydrophilicity	(Fujishima et al. 2000)
For photocatalytic energy conversion	(Tryk et al. 2000)
For surface patterning as a potential application of remote oxidation capability of TiO_2	(Tatsuma and Kubo 2007)

Table 1.3 (continued)

For details of basic photocatalytic mechanism	(Pelizzetti and Minero 1999)
On relative stability of TiO_2 semiconductor. Basics of photocatalysis	
Effect of operating variables and additives. Abatement of inorganic ions. Economic assessment of photocatalytic systems	(Gogate and Pandit 2004)
For purification against volatile organic compounds, kinetics, photocatalysts and reactors	(Zhao and Yang 2003)
For summary of the development regarding solar pilot plants, solar collectors and solar UV photocatalytic degradation	(Malato et al. 2002)
For highlights on spatially-confined photocatalytic systems	
Micro- and mesoporous photocatalysts; photocatalytic systems containing vesicles, emulsion systems, foams and layer-by- layer deposited polyelectrolyte capsules	(Shchukin and Sviridov 2006)
For basics of photocatalytic reactions. Photocatalytic reactions (e.g. photosynthesis, photodegradation, photofixation, photosensitisation) with examples	(Mills and Le Hunte 1997)
For review on photoelectrochemistry. Photocatalytic heterojuction systems. Optical properties of semiconductors vi-a-vis size. Goals for long term investigations of semiconductor photocatalytic remediation	(Levy 1997)
For nanosize TiO_2 powders versus nanothin films	(Hanley et al. 2009)
For synthesis, characterisation, electronic and computational studies of N-doped TiO_2	(Di Valentin et al. 2007)
For single visible light responsive, metal-, doped-, coupled, semiconductor photocatalysts. Factors determining photocatalytic properties	(Zhang et al. 2009)
For surface water and oxygen adsorption on semiconductor photocatalysts. Oxidation of aromatic and aliphatic compounds	
Reactor configurations	(Peral et al. 1997)
On membrane configurations for improving photocatalysis Catalyst recovery and recycle. Continuous stirred tank reactor (CSTR)	(Ollis 2003)
For definitions of terms under photocatalysis. Partial oxidationof alkanes	(Teichner 2008)
For use of photocatalysis in actinide valence control as potential waste management method	(Boxall 2005)
On TiO_2 photocatalytic oxidation. Types of photocatalytic reactors	
Membrane separation of TiO_2. Membrane hybrid systems	(Thiruvenkatachari et al. 2008)
For clear analysis of various topics under photocatalysis from first principles stating what is scientifically acceptable and what is possibly misconceived	(Ohtani 2010)
On photocatalysis using Cr(VI) and methylene blue as examples	
Visible light induced photocatalysis	(Rajeshwar 2001)
On photocatalysis, effect of operating parameters and chemical ions as additives	(Bhatkhande et al. 2001)
For protocols and terminologies in photocatalysis; photonic efficiency and quantum yield	(Serpone and Salinaro 1999)

Table 1.3 (continued)

On principles of application of semiconductor electrodes in photoelectrochemical cells; Photovoltaic, photoelectrosynthetic and photocatalytic applications	(Bard 1979)
On photocatalytic biocidal activity	(Banerjee et al. 2006)
For classification of areas under photocatalysis. Surface modification of TiO_2. Detailed discussion with specific examples	
Remote oxidation property of TiO_2	(Choi 2006)
For the determination of quantum yield	(Salinaro et al. 1999)
On the methods of improving photocatalysis by combination with other processes such as adsorption, fenton, ozonation, and sonication	(Augugliaro et al. 2006)
On solar applications, reactor design. Performance of different solar reactors and collectors and details of suitability of configuration for scale-up	(Braham and Harris 2009)
For metal oxide, sulphide and coupled photocatalytic systems	
Comparison with TiO_2	(Robert 2007)
For photocatalytic oxidation of aromatic and aliphatic compounds	
Photoctalytic reduction reactions of nitroaromatics, degradation and dehalogenation of chlorophenols, C-N coupling	(Shiraishi and Hirai 2008)
For primary processes involving charge carriers. Case-based review of photocatalytic treatment of waste water. Solar reactors	(Robertson et al. 2005)
For patented photocatalytic systems for air and water purification, and major manufacturers	(Mills and Lee 2002)
On the modification of TiO_2 towards stability, efficient light utilisation, dye mineralisation. Photocatalyst use and reuse	(Han et al. 2009)
For general concept of phoelectrochemical cells: photoelectrosynthesis, photocatalytic cells	(Bard 1979)
For photochemistry of TiO_2 surface, surface defects, charge carrier trapping and oxygen adsorption. N-doping	(Thompson and Yates 2005)
On solar photocatalysis for water treatment with practical examples	
Inactivation of microorganisms by TiO_2	(Blanco-Galvez et al. 2007)
For effect variables such as adsorption, temperature, ionic additives e.t.c. on photocatalysis. Listing of compounds degraded by photocatalysis, intermediates, catalysts and source of light used	(Bhatkhande et al. 2001)
For visible light-assisted photocatalytic degradation of compounds	
Modifications of TiO_2 for better performance	(Chatterjee and Dasgupta 2005)
For effect of transition metal ions on photocatalytic oxidation. Transformation of harmful transition metals. Photocatalytic metal deposition	(Litter 1999)
For electronic aspect of semiconductor photocatalysis.	
Chemisorption of gases on TiO_2. Photoaccelerated reactions of different types including water splitting	(Linsebigler et al. 1995)
For the photocatalytic dye removal. Influence of operating parameters and effect of oxidants, and analysis of products	(Konstantinou and Albanis 2004)
On the historical overview of photocatalysis.	

Table 1.3 (continued)

Photoelectrochemical basis. Properties of TiO_2. Applications of photocatalysis	(Fujishima et al. 2008)
For photocatalytic ozonation. Mechanism. Effect of ozone dosage	(Agustina et al. 2005)
For definition of photocatalysis, photoinitiated catalysis, photosensitisation e.t.c.	(Parmon 1997)
For general introduction of TiO_2 photocatalysis. The use of metal cocatalysts to improve photocatalytic oxidation	(Sobczyński and Dobosz 2001)
For photocatalytic degradation pathways. Overview on types of solar reactors and plants	(Bahnemann 2004)
On photocatalysis, mechanism and controversies on terminologies	(Serpone 1995)

1.8 Conclusion

The advent of semiconductor photocatalysis has highlighted several photodegradation phenomena which were previously unclearly understood. In fact, photocatalysis has been put forward to explain the non-detection of organic compounds in Mars based on the fact that there is presence of TiO_2, UV irradiation and O_2 in the Martian environment. At present, photocatalysis is dramatically drawing attention of scientific community and investors owing to its green attributes and cost effectiveness. Even though the fundamentals of photocatalysis regarding principle, reactions and elucidation of intermediate species have been well established, a few equivocal issues require further investigation particularly the mechanistic aspect, which includes the role of reactive oxidants in the semiconductor-mediated photocatalytic oxidation. Subsequent chapters will highlight on scientific responses and approaches followed to demystify these problems. Challenges for industrial application of photocatalytic synthesis as well as a few other obstacles still remain. Supposedly, near IR photocatalysis is a potential efficient method for future application in photocatalysis.

References

Abramović BF, Anderluh VB, Topalov AS et al (2004) Kinetics of photocatalytic removal of 2-amino-5-chloropyridine from water. APTEFF 35:1–280

Adesina AA (2004) Industrial exploitation of photocatalysis: progress, perspectives and prospects. Catal Surv Asia 8(4):265–273

Agustina TE, Ang HM, Vareek VK (2005) A review of synergistic effect of photocatalysis and ozonation on wastewater treatment. J Photochem Photobiol C 6:264–273

Alfano OM, Bahnemann D, Cassano AE et al (2000) Photocatalysis in water environments using artificial and solar light. Catal Today 58:199–230

Allaoui A, Malouki MA, Wong-Wah-Chung P (2010) Homogeneous photodegradation study of 2-mercaptobenzothiazole photocatalysed by sodium decatungstate salts: Kinetics and mechanistic pathways. J Photochem Photobiol A Chem 212:153–160

Amalric L, Guillard C, Pichat P et al (1995) Assessment of the importance of the role of H_2O_2 and bullet in the photocatalytic degradation of 1,2-dimethoxy benzene. Sol Energy Mater Sol Cells 38:391–399

Anpo M, Kubokawa Y (1984) Photoluminescence of zinc oxide powder as a probe of electron-hole surface processes. J Phys Chem 88:5556–5560

Argazzi R, Bignozzi CA (1995) Long-lived photoinduced charge separation across nanocrystalline TiO_2 interfaces. J Am Chem Soc 117:11815–11816

Augugliaro V, Litter M, Palmisano L et al (2006) The combination of heterogeneous photocatalysis with chemical and physical operations: a tool for improving the photoprocess performance. J Photochem Photobiol C 7:127–144

Auguliaro V, Davi E, Palmisano L et al (1990) Influence of hydrogen peroxide on the kinetics of phenol photodegradation in aqueous titanium dioxide dispersion. Appl Catal 65:101–116

Bahnemann D (2004) Photocatalytic water treatment: solar energy applications. Sol Energy 77:445–459

Bahnemann D, Henglein A, Lilie J et al (1984) Flash photolysis observation of the absorption spectra of trapped positive holes and electrons in colloidal TiO_2. J Phys Chem 88:709–711

Bahnemann DW, Hilgendorff M, Memming R (1997) Charge carrier dynamics at TiO_2 particles: reactivity of free and trapped holes. J Phys Chem B 101:4265–4275

Bahnemann DW, Dillert R, Robertson PKJ (2003) Photocatalysis: Initial reaction steps. In: Kokorin AI, Bahnemann DW (eds) Chemical physics of nanostructured semiconductors. VSP, Boston, pp 199–200

Balcerski W, Ryu SY, Hoffmann MR (2008) Gas-phase photodegradation of decane and methanol on TiO2: dynamic surface chemistry characterized by diffuse reflectance FTIR. Int J Photoenergy, 9 pages, Hindawi Publishing Corporation. doi:10.1155/2008/964721

Banerjee S, Gopal J, Muraleedharan P (2006) Physics and chemistry of photocatalytic titanium dioxide: visualization of bactericidal activity using atomic force microscopy. Curr Sci 90:1378–1383

Bard AJ (1979) Photoelectrochemistry and heterogeneous photocatalysis at semiconductors. J Photochem 10:59–75

Belgiorno V, Rizzo L, Fatta D et al (2007) Review on endocrine disrupting-emerging compounds in urban wastewater: occurrence and removal by photocatalysis and ultrasonic irradiation for wastewater reuse. Desalination 215:166–176

Bhatkhande DS, Pangarkar VG, Beenackers AACM (2001) Photocatalytic degradation for environmental applications—a review. J Chem Technol Biotechnol 77:102–116

Blake DM (1999) Bibliography of work on the heterogeneous photocatalytic removal of hazardous compounds from water and air, Update 3 to January 1999 (NREL/TP-570–26797). National Renewable Energy Laboratory, United States

Blanco J, Malato S, Fernández-Ibáez P et al (2009) Review of feasible solar energy applications to water processes. Renew Sustain Energy Rev 13:1437–1445

Blanco-Galvez J, Fernández-Ibáez P, Malato-Rodríguez S (2007) Solar photocatalytic detoxification and disinfection of water: recent overview. J Sol Energy Eng 129:4–15

Boxall C, Le Gurun G, Taylor RJ et al (2005) The applications of photocatalytic waste minimisation in nuclear fuel processing. In: Boule P et al (ed) Hdb Env Chem vol 2M. Springer-Verlag, Heidelberg, pp 451–481

Braham RJ, Harris AT (2009) Review of major design and scale-up considerations for solar photocatalytic reactors. Ind Eng Chem Res 48:8890–8905

Braslavsky SE (2007) International union of pure and applied chemistry, organic and biomolecular chemistry division, subcommittee on photochemistry glossary of terms used in photochemistry, 3rd ed. (IUPAC recommendations 2006). Pure Appl Chem 79(3):293–465

Calza P, Pelizzetti E, Minero C (2005) The fate of organic nitrogen in photocatalysis: an overview. J Appl Electrochem 35:665–673

Cameron RE, Bocarsly AB (1985) Photoactivated oxidation of alcohols by oxygen. J Am Chem Soc 107:6116–6117

Carmen W (2009) A comprehensive dictionary of inorganic chemistry. Abhishek Publication, India
Carp O, Huisman CL, Reller A (2004) Photoinduced reactivity of Titanium dioxide, Progress in solid state chemistry. Prog Solid State Chem 32:33–177
Castellan GW (ed) (1983) Physical chemistry. Addison-Wesley Publishing Company Inc, USA
Chatterjee D, Dasgupta S (2005) Visible light induced photocatalytic degradation of organic pollutants. J Photochem Photobiol C 6:186–205
Choi W (2006) Pure and modified TiO_2 photocatalysts and their environmental applications. Catal Surv Asia 10:16–28
Chu W, Wong CC (2004) The photocatalytic degradation of dicamba in TiO_2 suspensions with the help of hydrogen peroxide by different near UV irradiations. Water Res 38:1037–1043
Colón G, Hidalgo MC, Navío JA (2001) Influence of carboxylic acid on the photocatalytic reduction of Cr(VI) using commercial TiO_2. Langmuir 17:7174–7177
Daimon T, Nosaka Y (2007) Formation and behavior of singlet molecular oxygen in TiO_2 photocatalysis studied by detection of near-infrared phosphorescence. J Phys Chem C 111:4420–4424
Daimon T, Hirakawa T, Kitazawa M et al (2008) Formation of singlet molecular oxygen associated with the formation of superoxide radicals in aqueous suspensions of TiO_2 photocatalysts. Appl Catal A 340:169–175
David T (1995) Insights into specialty inorganic chemicals. Royal Society of Chemistry, UK
De Lasa H, Serrano B, Salaices M (2005) Photocatalytic Reaction Engineering. Springer, New York
Desilvestro J, Gratzel M, Kavan L et al (1985) Highly efficient sensitization of titanium dioxide. J Am Chem Soc 107:2988–2990
Di Valentin C, Finazzi E, Pacchioni G et al (2007) N-doped TiO_2: theory and experiment. Chem Phys 339:44–56
Dillert R, Cassano AE, Goslich R et al (1999) Large scale studies in solar catalytic wastewater treatment. Catal Today 54:267–282
Doong R-A, Maithreepala R-A, Chang S-M (2000) Heterogeneous and homogeneous photocatalytic degradation of chlorophenols in aqueous titanium dioxide and ferrous ion. Water Sci Technol 42(7–8):253–260
Draper RB, Fox MA (1990) Titanium dioxide photosensitized reactions studied by diffuse reflectance flash photolysis in aqueous suspensions of TiO_2 powder. Langmuir 6:1396–1402
Du Y, Rabani J (2003) The measure of TiO_2 photocatalytic efficiency and the comparison of different photocatalytic titania. J Phys Chem B 107:11970–11978
Eiovitz MS, Fish W (1994) Redox interactions of Cr(VI) and substituted phenols: Kinetic investigation. Environ Sci Technol 28:2161–2169
EPA (1998) Superfund, Record of decision: Ohio river park, EPA ID: PAD980508816 OU 03. Environmental protection agency, Neville Island. United States of America, Pennsylvania, pp 1–33
Fan JC, Sreekanth KM, Xie Z et al (2013) p-Type ZnO materials: theory, growth, properties and devices. Prog Mater Sci 58:874–985
Feng J, Han J, Zhao X (2009) Synthesis of $CuInS_2$ quantum dots on TiO_2 porous films by solvothermal method for absorption layer of solar cells. Prog Org Coat 64:268–273
Fox MA, Chen M-J (1983) Photocatalytic formylation of primary and secondary amines on irradiated semiconductor powders. J Am Chem Soc 105:4497–4499
Fox MA, Ogawa H, Pichat P (1989) Regioselectivity in the semiconductor-mediated photooxidation of 1,4-pentanediol. J Org Chem 54:3847–3852
Fu H, Zhang L, Zhang S et al (2006) Electron spin resonance spin-trapping detection of radical intermediates in N-doped TiO_2-assisted photodegradation of 4-chlorophenol. J Phys Chem B 110:3061–3065
Fujishima A, Honda K (1972) Electrochemical photolysis of water at a semiconductor electrode. Nature 238:37–38 (London, UK)
Fujishima A, Rao TN, Tryk DA (2000) Titanium dioxide photocatalysis. J Photochem Photobiol C 1:1–21

Fujishima A, Zhang X, Tryk DA (2007) Heterogeneous photocatalysis: from water photolysis to applications in environmental cleanup. Int J Hydrogen Energy 32:2664–2672
Fujishima A, Zhang X, Tryk DA (2008) TiO_2 photocatalysis and related surface phenomena. Surf Sci Rep 63:515–582
Gao R, Stark J, Bahnemann DW et al (2002) Quantum yields of hydroxyl radicals in illuminated TiO_2 nanocrystallite layers. J Photochem Photobiol A 148:387–391
Gaya UI (2011) Comparative analysis of ZnO−catalyzed photo−oxidation of *p*−chlorophenols. Eur J Chem 2:163−167
Gogate PR, Pandit AB (2004) A review of imperative technologies for wastewater treatment I: oxidation technologies at ambient conditions. Adv Environ Res 8:501–551
Goodeve CF, Kitchener JA (1938) Photosensitization by titanium dioxide. Trans Faraday Soc 34:570–579
Hagen J (2006) Industrial catalysis. Wiley-VCH Verlag GmbH & Co. KGaA, Germany
Haick H, Paz Y (2003) Long-range effects of noble metals on the photocatalytic properties of titanium dioxide. J Phys Chem B 107:2319–2326
Han F, Kambala VSR, Srinivasan M et al (2009) Tailored titanium dioxide photocatalysts for the degradation of organic dyes in wastewater treatment: a review. Appl Catal A 359:25–40
Hanley T, Krisnandi Y, Eldewik A (2001) Nanosize effects in titania based photocatalyst materials. Ionics 7:319
Harbour JR, Tromp J, Hair ML (1985) Photogeneration of hydrogen peroxide in aqueous TiO_2 dispersions. Can J Chem 63:204–208
Hashimoto K, Irie H, Fujishima A (2005) TiO_2 Photocatalysis: A overview and future prospects. Jpn J Appl Phys 44(12):8269–8285
Herrmann J-M (1999) Heterogeneous photocatalysis: fundamentals and applications to the removal of various types of aqueous pollutants. Catal Today 53:115–129
Herrmann J-M (2005) Heterogeneous photocatalysis: state of the art and present applications. Top Catal 34:1–4
Herrmann J-M (2010a) Fundamentals and misconceptions in photocatalysis. J Photochem Photobiol A 216:85–93
Herrmann J-M (2010b) Photocatalysis fundamentals revisited to avoid several misconceptions. Appl Catal B 99:461–468
Herrmann J-M, Duchamp C, Karkmaz M et al (2007) Environmental green chemistry as defined by photocatalysis. J Hazard Mater 146:624–629
Hirakawa T, Nosaka Y (2002) Properties of $O_2^{\bullet -}$ and OH^- formed in TiO_2 aqueous suspensions by photocatalytic reaction and the influence of H_2O_2 and some ions. Langmuir 18:3247–3254
Hoffman AJ, Carraway ER, Hoffmann MR (1994) Photocatalytic Production of H_2O_2 and organic peroxides on quantum-sized semiconductor collolds. Environ Sci Technol 28:776–785
Hoffmann MR, Martin ST, Choi W et al (1995) Environmental applications of semiconductor photocatalysis. Chem Rev 95:69–96
Horváth IT (2003) Encyclopedia of catalysis. Wiley, New Jersey
Housecroft CE, Sharfe AG (eds) (2005) Inorganic chemistry. Pearson Education Limited, England
Iliev V (2002) Phthalocyanine-modified titania—catalyst for photooxidation of phenols by irradiation with visible light. J Photochem Photobiol A Chem 151:195–199
Ilisz I, Föglein K, Dombi A (1998) The photochemical behavior of hydrogen peroxide in near UV-irradiated aqueous TiO_2 suspensions. J Mol Catal A Chem 135:55–61
Imoberdorf GE, Irazoqui HA, Cassano AE et al (2005) Photocatalytic degradation of tetrachloroethylene in gas phase on TiO_2 films: A kinetic study. Ind Eng Chem Res 44:6075–6085
INERIS (2005) Institut National de l'environnement industriel et de risqué (INERIS). http://www.ineris.fr/. Accessed 19 May 2013
Ishibashi K, Fujishima A, Watanabe T et al (2000) Quantum yields of active oxidative species formed on TiO_2 photocatalyst. J Photochem Photobiol A 134:139–142
Jańczyk A, Krakowska E, Stochel G et al (2006) Singlet oxygen photogeneration at surface modified titanium dioxide. J Am Chem Soc 128:15574–15575

Jiuhui QU (2008) Research progress of novel adsorption processes in water purification: a review. J Environ Sci 20:1–13
Kabra K, Chaudhary R, Sawhney RL (2004) Treatment of hazardous organic and inorganic compounds through aqueous-phase photocatalysis: a review. Ind Eng Chem Res 43:7683–7696
Kako T, Kikugawa N, Ye J (2008) Photocatalytic activities of $AgSbO_3$ under visible light irradiation. J Catal Today 131:197–202
Kato S, Mashio F (1956) Autooxidation by TiO_2 as a photocatalyst. Abtr Book Annu Meet Chemical Society of Japan 223
Khan AU (1970) Singlet molecular oxygen from superoxide anion and sensitized fluorescence of organic molecules. Science 168:476–477
Kluson P, Drobek M, Strasak T et al (2007) Sulphonated phthalocyanines as effective oxidation photocatalysts for visible and UV light regions. J Mol Catal A Chem 272:213–219
Konstantinou IK, Albanis TA (2004) TiO_2-assisted photocatalytic degradation of azo dyes in aqueous solution: kinetic and mechanistic investigations. Appl Catal B 49:1–14
Levy B (1997) Photochemistry of nanostructured materials for energy applications. J of Electroceram 1(3):239–272
Lewis RJ Sr (ed) (2002) Hazardous chemicals desk reference. Wiley, New York
Linsebigler AL, Lu G, Yates JTJetal (1995) Photocatalysis on TiO_2 surfaces: principles, mechanisms, and selected results. Chem Rev 95:735–758
Litter MI (1999) Heterogeneous photocatalysis. Transition metal ions in photocatalytic systems. Appl Catal B 23:89–114
Litter MI (2005) Introduction to photochemical advanced oxidation processes for water treatment. In: Boule P, Bahnemann DW, Robertson PKJ (eds) Hdb Env Chem Vol 2M. Springer-Verlag, Heidelberg, pp 325–366
Lu G, Linsebigler A, Yates JT Jr (1995) Photooxidation of CH_3Cl on TiO_2 (110): a mechanism not involving H_2O. J Phys Chem 99:7626–7631
Malato S, Blanco J, Vidal A et al (2002) Photocatalysis with solar energy at a pilot-plant scale: an overview. Appl Catal B 37:1–15
Matthews RW (1992) Photocatalytic oxidation of organic contaminants in water: an aid to environmental preservation. Pure Appl Chem 64(9):1285–1290
Mattioli G, Filippone F, Alippi P et al (2008) Ab initio study of the electronic states induced by oxygen vacancies in rutile and anatase TiO_2. Phys Rev B 78:241201-1-241201-4
Mele G, Sole RD, Vasapollo G et al (2003) Photocatalytic degradation of 4-nitrophenol in aqueous suspension by using polycrystalline TiO_2 impregnated with functionalized Cu(II)–porphyrin or Cu(II)–phthalocyanine. J Catal 217:334–342
Mills A, Le Hunte S (1997) An overview of semiconductor photocatalysis. J Photochem Photobiol A 108:1–35
Mills A, Lee S-K (2002) A web-based overview of semiconductor photochemistry-based current commercial applications. J Photochem Photobiol A 152:233–247
Minero C, Mariella G, Maurion V et al (2000) Photocatalytic transformation of organic compounds in the presence of inorganic anions. 2. Competitive reactions of phenol and alcohols on a titanium dioxide-fluoride system. Langmuir 16:8964–8972
Morrison SR, Freund T (1967) Chemical role of holes and electrons in ZnO photocatalysis. J Chem Phys 47(4):1543–1541
Mrowetz M, Selli E (2006) H_2O_2 evolution during the photocatalytic degradation of organic molecules on fluorinated TiO_2. New J Chem 30:108–114
Muruganandham M, Sobana N, Swaminathan M (2006) Solar assisted photocatalytic and photochemical degradation of Reactive Black 5. J Hazard Mater B 137:1371–1376
Naman SA, Khammas ZA-A, Hussein FM (2002) Photo-oxidative degradation of insecticide dichlorovos by a combined semiconductors and organic sensitizers in aqueous media. J Photochem Photobiol A 153:229–236
Nowotny MK, Sheppard LR, Bak T et al (2008) Defect chemistry of titanium dioxide. application of defect engineering in processing of TiO_2-based photocatalysts. J Phys Chem C 112:5275–5300

Ohtani B (2008) Preparing articles on photocatalysis—beyond the illusions, misconceptions, and speculation. Chem Lett 37:217–229

Ohtani B (2010) Photocatalysis A to Z—what we know and what we do not know in a scientific sense. J Photochem Photobiol C 11:157–178

Ohuchi T, Miyatake T, Hitomi Y et al (2007) Liquid phase photooxidation of alcohol over niobium oxide without solvents. Catal Today 120:233–239

Ollis DF (2000) Photocatalytic purification and remediation of contaminated air and water. C R Acad Sci Paris Série IIc Chimie Chem 3:405–411

Ollis DF (2003) Integrating photocatalysis and membrane technologies for water treatment. Ann NY Acad Sci 984:65–84

Ozawa K, Sawada K, Shirotori Y et al (2005) Angle-resolved photoemission study of the valence band structure of ZnO (10 10). J Phys Condens Matter 17:1271–1278

Parmon VN (1997) Photocatalysis as a phenomenon: aspect of terminology. Catal Today 39:137–144

Patnaik P (ed) (2004) Dean's analytical chemistry handbook. McGraw-Hill Handbooks, New York

Pelizetti E, Maurino V, Minero C et al (1990) Photocatalytic degradation of atrazine and other s-triazine herbicides. Environ Sci Technol 24:1559–1565

Pelizzetti E, Minero C (1999) Mechanism of the photo-oxidative degradation of organic pollutants over TiO_2 particles. Electrochim Acta 38:47–55

Peral J, Doménech X, Ollis DF (1997) Heterogeneous photocatalysis for purification, decontamination and deodorization of air. J Chem Technol Biotechnol 70:117–140

Peter FS, Nicholas JT, Stefan HB et al (1997) A new method to determine the generation of hydroxyl radicals in Illuminated TiO_2 Suspensions. J Phys Chem B 101:7127–7134

Pohanish RP, Greene SA (1996) Hazardous material handbook. Van Nostrand Reinhold/International Thompson Publishing, New York

Rajeshwar K, Chenthamarakshan CR, Goeringer S et al (2001) Titania-based heterogeneous photocatalysis. Materials, mechanistic issues, and implications for environmental remediation. Pure Appl Chem 73(12):1849–1860

Rajeshwar K, Osugi ME, Chanmanee W et al (2008) Heterogeneous photocatalytic treatment of organic dyes in air and aqueous media. J Photochem Photobiol C 9:171–192

Rajh T, Poluektov OG, Thurnauer MC (2003) Charge separation in titanium oxide nanocrystalline semiconductors revealed by magnetic resonance. In: Kokorin AI, Bahnemann DW (eds) Chemical physics of nanostructured semiconductors. VSP, Boston, pp 1–34

Ranjit KT, Viswanathan B (1997) Synthesis, characterization and phtocatalytic properties of iron-doped TiO catalysts. J Photochem Photobiol A 108:79–84

Ravichandran L, Selvam K, Swaminathan M (2007) Effect of oxidants and metal ions on photodefluoridation of pentafluorobenzoic acid with ZnO. Sep Purif Technol 56:192–198

Renz C (1921) Photoreactions of oxides of titanium, cerium and earth acids. Helv Chim Acta 4:961–968

Richard C (1993) Regioselectivity of oxidation by positive holes (h^+) in photocatalytic aqueous transformations. J Photochem Photobiol A 72:179–182

Richard CN (ed) (1995) Solar energy conversion. The solar cell. Elsevier

Richard C, Boule P (1991) Oxidising species involved intransfortions on zinc oxide. J Photochem Photobiol A 60:235–243

Robert D (2007) Photosensitization of TiO_2 by M_xO_y and M_xS_y nanoparticles for heterogeneous photocatalysis applications. Catal Today 122:20–26

Robertson PKJ, Bahnemann DW, Robertson JMC et al (2005) Photocatalytic detoxification of water and air. In: Boule P, Bahnemann DW, Robertson PKJ (eds) Hdb Env Chem, vol 2. Springer-Verlag, Heidelberg, pp 367–423

Rodríguez JA, Fernández-García M (2007) Synthesis, properties and applications of oxide nanomaterials. Wiley, New Jersey

Ropp RC (2003) Solid state chemistry. Elsevier B.V., Amsterdam

Saien J, Soleymani AR (2007) Degradation and mineralization of direct blue 71 in a circulating upflow reactor by UV/TiO_2 process and employing a new method in kinetic study. J Hazard Mater 144:506–512
Salinaro A, Emeline AV, Zhao J et al (1999) Terminology, relative photonic efficiencies and quantum yields in heterogeneous photocatalysis. Part I: experimental determination of quantum yield. Pure Appl Chem 71:321–335
Salvador P (2007) On the nature of photogenerated radical species active in the oxidative degradation of dissolved pollutants with TiO_2 aqueous suspensions: a revision in the light of the electronic structure of adsorbed water. J Phys Chem C 111:17038–17043
Saxena SC, Jotshi CK (1996) Management and combustion of hazardous waste. Prog Energy Combust Sci 22:401–425
Sclafani A, Herrmann J-M (1998) Influence of metallic silver and of platinum-silver bimetallic deposits on the photocatalytic activity of titania (anatase and rutile) in organic and aqueous media. J Photochem Photobiol A 113:181–188
Sehili T, Boule P, Lemaire J (1991) Photocatalysed transfornation of chloroaromatic derivatives on zinc oxide IV: 2,4-dichlorophenol. Chemosphere 22:1053–1062
Serpone N (1995) Brief Introductory remarks on heterogeneous photocatalysis. Sol Energy Mater Sol Cells 38:369–379
Serpone N, Khairutdinov RF (1996) Application of nanoparticles in the photocatalytic degradation of water pollutants. In: Kamat PV, Meisel (eds) Semiconductor nanoclusters. Studies in surface science and catalysis, vol 103. Elsevier Science, pp 417–444
Serpone N, Pelizzetti E (1989) Photocatalysis, fundamentals and applications. Wiley, New York
Serpone N, Salinaro A (1999) Terminology, relative photonic efficiencies and quantum yields in heterogeneous photocatalysis. Part I: suggested protocol. Pure Appl Chem 71(2):303–320
Shchukin DG, Sviridov DV (2006) Photocatalytic processes in spatially confined micro- and nanoreactors. J Photochem Photobiol C 7:23–39
Shiraishi Y, Hirai T (2008) Selective organic transformations on titanium oxide-based photocatalysts. J Photochem Photobiol C 9:157–170
Shiying Y, Ping W, Xin Y et al (2009) A novel advanced oxidation process to degrade organic pollutants in wastewater: microwave-activated persulfate oxidation. J Environ Sci 21:1175–1180
Smart LE, Moore EA (2005) Solid state chemistry: an introduction. Taylor & Francis group, Boca Raton.
Sobczyński A, Dobosz A (2001) Water purification by photocatalysis on semiconductors. Pol J Environ Stud 10(4):195–205
Stafford U, Gray KA, Kamat PV (1994) Radiolytic and TiO_2-assisted photocatalytic degradation of 4-chlorophenol: a comparative study. J Phys Chem 98:6343–6351
Sun B, Smirniotis PG, Boolchand P (2005) Visible light photocatalysis with platinized rutile TiO_2 for aqueous organic oxidation. Langmuir 21:11397–11403
Swaddle TW (1990) Inorganic chemistry. Academic Press, USA
Tachibana Y, Moser JE, Grätzel M et al (1996) Subpicosecond interfacial charge separation in dye-sensitized nanocrystalline titanium dioxide films. J Phys Chem 100:20056–20062
Tai C, Jiang G (2005) Dechlorination and destruction of 2,4,6-trichlorophenol and pentachlorophenol using hydrogen peroxide as the oxidant catalyzed by molybdate ions under basic condition. Chemosphere 59:321–326
Tatsuma T, Kubo W (2007) Photocatalytic lithography based on photocatalytic remote oxidation. J Photopolym Sci Technol 20:83–86
Tauc J, Menth AJ (1972) States in the gap. Non-Cryst Solids 8–10:569–585
Teichner SJ (2008) The origins of photocatalysis. J Porous Mater 15:311–314
Thiruvenkatachari R, Vigneswaran S, Moon IS (2008) A review on UV/TiO_2 photocatalytic oxidation process. Korean J Chem Eng 25:64–72
Thompson TL, Yates JT Jr (2005) TiO_2-based photocatalysis: surface defects, oxygen and charge transfer. Top Catal 35:197–210

Tompkins DT, Anderson MA (2001) Evaluation of photocatalytic air cleaning capability. Final report, ASHRAE 1134-RP, pp 1–93

Trikalitis PN, Rangan KK, Bakas T et al (2001) Varied pore organization in mesostructured semiconductors based on the $[SnSe_4]^{4-}$ anion. Nature 410:671–674

Tryk DA, Fujishima A, Honda K (2000) Recent topics in photoelectrochemistry: achievements and future prospects. Electrochim Acta 45:2363–2376

Turchi CS, Ollis DF (1990) Photocatalytic degradation of organic water contaminants: mechanisms involving hydroxyl radical attack. J Catal 122:178–192

US EPA (2006) National emission standards for hazardous air pollutants: miscellaneous organic chemical manufacturing; Final Rule, Federal Register Part V, 40 CFR Part 63, United States Environmental Protection Agency, pp 40316–40342

US EPA/OSHA (2000) Chemical advisory and notice of potential risk: Skin exposure to molten 2,4-dichlorophenol (2,4-DCP) can cause rapid death, US EPA Office of pollution prevention and toxics and occupational safety and health administration, United States of America, pp 1–13

Wang C-Y, Pagel R, Dohrmann JK et al (2006) Antenna mechanism and deaggregation concept: novel mechanistic principles for photocatalysis. C R Chimie 9:761–773

Wang J, Tafen DN, Lewis JP et al (2009) Origin of photocatalytic activity of nitrogen-doped TiO_2 nanobelts. J Am Chem Soc 131:12290–12297

Wang S, Wang Z, Zhuang Q (1992) Photocatalytic reduction of the environmental pollutant Cr(VI) over CdS powder under visible illumination. Appl Catal B 1:257–270

Wang X, Pehkonen SO, Ray AK (2004) Photocatalytic reduction of Hg(II) on two commercial TiO_2 catalysts. Electrochim Acta 49:1435–1444

Wang Z-H, Zhuang Q-X (1993) Photocatalytic reduction of pollutant Hg(II) on WO_3 dispersion. J Photochem Photobiol A Chem 75:105–111

Ward MD, Bard AJ (1982) Photocurrent enhancement via trapping of photogenerated electrons of TiO_2 Particles. J Phys Chem 86:3599–3805

Willie G (2010) A comprehensive dictionary of chemistry. Abhishek Publication, India

Wu L, Li A, Gao G et al (2007) Efficient photodegradation of 2,4-dichlorophenol in aqueous solution catalyzed by polydivinylbenzene-supported zinc phthalocyanine. J Mol Catal A Chem 269:183–189

Yamazoe S, Okumura T, Hitomi Y et al (2007) Mechanism of Photo-Oxidation of NH_3 over TiO_2: fourier transform infrared study of the intermediate species. J Phys Chem C 111:11077–11085

Yang L, Yu LE, Ray MB (2009) Photocatalytic oxidation of paracetamol: dominant reactants, intermediates, and reaction mechanisms. Environ Sci Technol 43:460–465

Yoon SH, Lee JH, Oh SE et al (2008) Photochemical oxidation of As(III) by vacuum-UV lamp irradiation. Water Res 42:3455–3463

Zhao J, Yang X (2003) Photocatalytic oxidation for indoor air purification: a literature review. Build Environ 38:645–654

Zhang D, Wu L-Z, Yang Q-Z et al (2003) Versatile photosensitization system for TiO_2-mediated oxidation of alkenes based on nafion-supported platinum(II) terpyridyl acetylide complex. Org Lett 5:3221–3224

Zhang H, Chen G, Bahnemann DW (2009) Photoelectrocatalytic materials for environmental applications. J Mater Chem 19:5089–5121

Chapter 2
Kinetic Concepts of Heterogeneous Photocatalysis

Abstract The ABCs of empirical chemical kinetics are highlighted from the scratch. Aspects of dynamics regarding charge carrier trapping and recombination are essential part of this discussion. The effect of imposition of external constrains such as temperature of the reactivity solution, concentration of the subject compound, intensity and wavelength of light on photocatalytic degradation system is emphasised. Because of the crucial position of light intensity in heterogeneous photocatalysis, the principles of operation of important chemical actinometric systems are presented.

2.1 Background Sketch

Chemical kinetics addresses important issues in heterogeneous photocatalysis. Generally, it deals with the experimental determination and analysis of the quantity of a substance (usually concentration) as a function of time. In other words, chemical kinetics is concerned with reaction rate and throughput. Unlike in chemical equilibrium where the reactants and the products are in dynamic state of balance, chemical kinetics is concerned with either disappearance or appearance of chemical substances. The latter have direct consequences on reaction rate. A given chemical reaction may get accelerated or slowed down, for example by a positive catalyst or negative catalyst respectively as shown in Fig. 2.1. A simple example of this concept can be derived from common domestic techniques such as preservation using refrigeration method. The shell life of batteries can be extended by lowering the surrounding temperature. The simple reason behind this observation is that the rate of decay of the electrolytes or sensitisers in the battery is slowed down by lowering the temperature. The following discussions will cover aspects of chemical kinetics including the factors that affect the rate of photocatalytic reactions.

Some aspects of catalysis such as the determination of rate are viewed from the standpoint of the principles of chemical kinetics. The rate of reaction is measured based on the property that is easiest to measure. This property can be colour, concentration or weight. As several products of varied structure and quantity can be formed during photocatalytic reactions, it is more convenient to measure reaction rates based on the disappearance of such a property.

The law of mass action is the central to the expression of reaction rates. The law is based on the fact that the rate of reaction increases proportionally with concentration.

U. I. Gaya, *Heterogeneous Photocatalysis Using Inorganic Semiconductor Solids*, DOI 10.1007/978-94-007-7775-0_2,

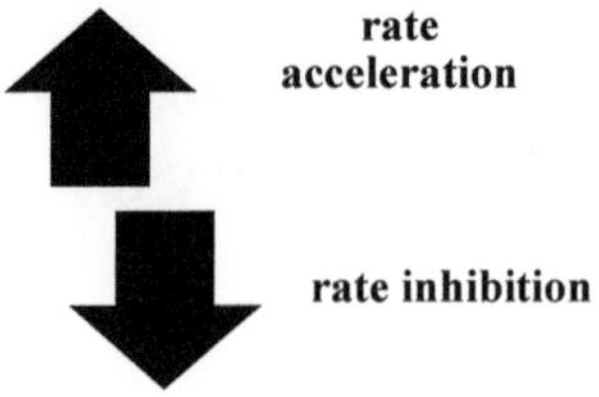

Fig. 2.1 Types of kinetic influences. Processes can be manipulated using a suitable catalyst to decline or increase in reaction rate, depending on the type of catalyst or reaction condition

This direct variation is expressed mathematically as a rate law and the constant of proportionality is termed rate constant.

Consider the photocatalytic transformation of a molecule A:

$$\mathrm{A} \rightarrow \mathrm{products} \tag{2.1}$$

The rate of disappearance of A can be given by differential rate equation known as the differential rate law. A general differential rate law is given below.

$$\mathrm{r} = \frac{-\mathrm{d}\,[\mathrm{A}]}{\mathrm{dt}} = k[\mathrm{reactant}]^{\mathrm{n}} \tag{2.2}$$

where k is the rate constant and the power n is the order. A reaction is first, second or third order depending on the value of n (0, 1, 2 or 3 respectively). The parameter k as well can be first order, second order or third order rate constant. Rate constant can be utilised in comparing the rate of different reactions.

If time is measured in seconds the unit of r will be judged from Eq. (2.1) to be mol $\mathrm{dm^{-3}\ s^{-1}}$. The unit of k however varies with the reaction scheme. For zero order the rate law is

$$\mathrm{r}\,(\mathrm{mol\,dm^{-3}s^{-1}}) = k[\mathrm{A}]^0, \quad \text{or} \quad \mathrm{r} = \mathrm{k}$$

So for zero order reactions k is expressed in mol $\mathrm{dm^{-3}\ s^{-1}}$

Similarly, first order rate constant can be given by

$$k = \frac{\mathrm{r}\,(\mathrm{mol\ dm^{-3}s^{-1}})}{[\mathrm{A}](\mathrm{mol\ dm^{-3}})} \tag{2.3}$$

So, the unit of first order rate constant is $\mathrm{s^{-1}}$.

For second order reactions

$$k = \frac{\mathrm{r}\,(\mathrm{mol\ dm^{-3}s^{-1}})}{[\mathrm{A}]^2(\mathrm{mol\ dm^{-3}})^2} \tag{2.4}$$

Therefore the second order rate constant is expressed in $\mathrm{dm^3mol^{-1}\ s^{-1}}$. In this case, doubling the concentration of A results in quadrupling the rate of reaction.

For third order reactions,

$$k = \frac{\mathrm{r}\,(\mathrm{mol\ dm^{-3}s^{-1}})}{[\mathrm{A}]^3(\mathrm{mol\ dm^{-3}})^3} \tag{2.5}$$

So the third order rate constant is expressed in $dm^6\ mol^{-1}\ s^{-1}$. Since time is usually recorded in minutes during experiments many workers prefer to report rate constants with time in min^{-1}. Similarly, concentrations of substances with unknown molar masses have often been reported in ppm or $mgdm^{-3}$.

2.2 Decay Models in Photocatalysis

Various photocatalytic processes have been described by different kinetic models such as pseudo zero order, pseudo first order and second order models. As most of the processes studied are monitored based on the disappearance of the initial entities we refer to their kinetic schemes as decay models. The integral treatment of differential rate equations yields integrated rate laws which can display vividly the disappearance or decay profile of the original compound undergoing photocatalytic destruction. The commonest decay models that are observed in photocatalysis studies are the zero and the first order rate laws. Normally, the zero order is observed at surface saturation. Basically, the rate of a zero order reaction does not vary with concentration. In other words, a change in concentration does not result in any change in the reaction rate. In this case, the differential rate law for the disappearance of any reactant molecule A is given by

$$\frac{d[\mathrm{A}]}{dt} = -k \tag{2.6}$$

So

$$\mathrm{d}\,[\mathrm{A}] = -k\mathrm{dt} \tag{2.7}$$

Upon integration between the concentration limits $[A]_0$ and $[A]_t$ and time $t=0$ and $t=t$ we have

$$\int_{[\mathrm{A}]_0}^{[\mathrm{A}]_t} \mathrm{d}\,[\mathrm{A}] = k \int_{t=0}^{t=t} \mathrm{dt} \tag{2.8}$$

$$[\mathrm{A}]_t - [\mathrm{A}]_o = -k\mathrm{t} \tag{2.9}$$

Or

$$[\mathrm{A}]_t = [\mathrm{A}]_o - k\mathrm{t} \tag{2.10}$$

If Eq. (2.10) is compared with the equation of straight line $y = mx + c$, a plot of $[A]_t$ against t will give a straight line whose intercept equals $[A]_0$ and slope equals $-k$ (Fig. 2.2).

Conversely, the rate of first order reaction depends upon the concentration of single reactant. The rate of consumption of A is

$$-\frac{\mathrm{d}[\mathrm{A}]}{\mathrm{dt}} = k[\mathrm{A}] \tag{2.11}$$

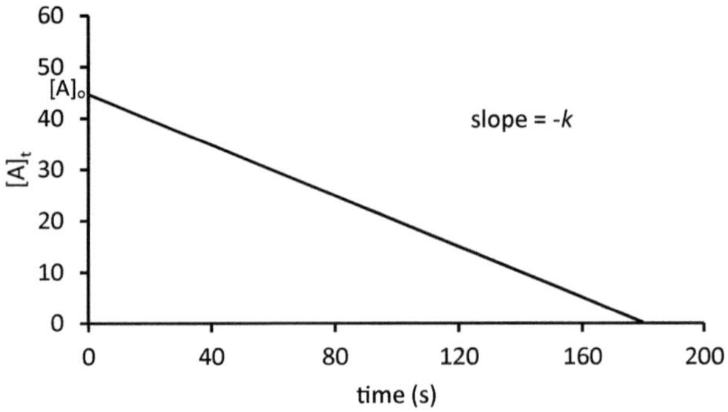

Fig. 2.2 Zero order kinetic fit for the disappearance of a reactant A with time

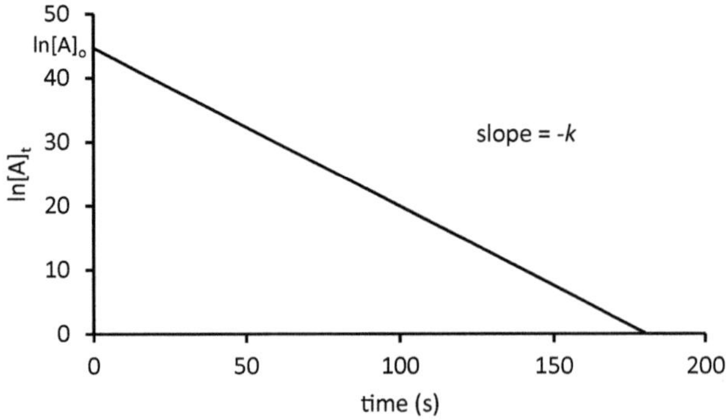

Fig. 2.3 First order fit for the decay of reactant A with concentration varying in the range $[A]_0$ to $[A]_t$ and with time up to t seconds

By collecting concentration terms together we obtain the following:

$$\frac{d[A]}{[A]} = -k\text{dt} \quad (2.12)$$

If the concentration of A at time t = 0 s is $[A]_0$ and at any other time t = t s the concentration is $[A]_t$ the differential rate law can be integrated within those limits as follows:

$$\int_{[A]_0}^{[A]_t} \frac{d[A]}{[A]} = k \int_{t=0}^{t=t} \text{dt} \quad (2.13)$$

$$\therefore \ln[A]_t - \ln[A]_0 = -kt \quad (2.14)$$

or

$$\ln[A]_t = \ln[A]_0 - kt \quad (2.15)$$

If Eq. (2.15) is compared with the equation of straight line $y = mx + c$, a plot of $\ln[A]_t$ versus t should give a straight line whose slope is $-k'$. The intercept of the plot is $\ln[A]_0$ (Fig. 2.3).

Eq. (2.15) can be presented in exponential form, to display the relationship between first order concentration decay and time.

$$\frac{[\mathrm{A}]_\mathrm{t}}{[\mathrm{A}]_0} = e^{-kt} \tag{2.16}$$

This decay equation has found important applications in physical chemistry which include the dating of prehistoric lives.

2.3 Pseudo Order Reactions

The term 'pseudo-' has been commonly used to prefix the order of a reaction because in real fact one of the reactants in catalytic systems acts as a catalyst and its concentration remains more or less unchanged such that instead of having second-order reaction for example, a pseudo-first order reaction is observed (Auguliaro et al. 2008). Similarly when the substrate is in large excess, the catalyst surface gets wholly covered all through and the disappearance of the substrate becomes negligibly dependent on concentration. At such saturation coverage, a pseudo zero order photocatalytic reaction will be observed (Theurich et al. 1996).

If a substance A is undergoes transformation such that its apparent concentration remains constant during the process, the following expression holds true.

$$\frac{d[\mathrm{A}]}{dt} = -k[\mathrm{A}] \tag{2.17}$$

Since [A] is constant the rate changes by a constant value which can be represented by

$$-\frac{\mathrm{d[A]}}{\mathrm{dt}} = -k' \tag{2.18}$$

$$\mathrm{d[A]} = -k'\mathrm{dt} \tag{2.19}$$

where k' is the observed or pseudo-zero order rate constant. If the concentration of A measured at time t = 0 s is $[\mathrm{A}]_0$ and any time t = t s the concentration becomes $[\mathrm{A}]_\mathrm{t}$ the differential rate law can be integrated within those limits as follows:

$$\int_{[\mathrm{A}]_0}^{[\mathrm{A}]_\mathrm{t}} \mathrm{d[A]} = -k' \int_{\mathrm{t}=0}^{\mathrm{t}=\mathrm{t}} \mathrm{dt} \tag{2.20}$$

$$[\mathrm{A}]_\mathrm{t} = [\mathrm{A}]_0 - k't \tag{2.21}$$

On the other hand, if we consider the two reacting species A with B combining to form products

$$\mathrm{A} + \mathrm{B} \rightarrow \mathrm{Products} \tag{2.22}$$

If the concentration of A and B at the start of the reaction be [A] and [B]. Then,

$$-\frac{d[A]}{dt} = k[A][B] \tag{2.23}$$

Assuming that [B] ≫ [A] or in other words [B] is constant then

$$-\frac{d[A]}{dt} = k'[A] \tag{2.24}$$

k' is the observed rate constant known as the pseudo-first order rate constant. Therefore, the major difference between pseudo first order rate equation and first order lies in the observed rate constant. If the concentration of A measured at time t = 0 s is $[A]_0$ and any time t = t s the concentration becomes $[A]_t$ the differential rate law can be integrated within those limits as follows:

$$\int_{[A]_o}^{[A]_t} \frac{d[A]}{[A]} = k' \int_{t=0}^{t=t} dt \tag{2.25}$$

$$\ln[A]_t - \ln[A]_0 = -k't \tag{2.26}$$

$$\ln[A]_t = \ln[A]_0 - k't \tag{2.27}$$

If Eq. (2.27) is compared with the equation of straight line y = mx + c, a plot of $\ln[A]_t$ versus t should give a straight line whose slope is $-k'$. The intercept of the plot is $\ln[A]_0$.

2.4 Fractional Lifetimes

Within certain time of the start of chemical reaction, only a fraction of the original substance may be transformed. The time required for this transformation is known as fractional lifetime. The time taken for the original amount of reactant to transform is called the half-life. Half-life has been the conventional fractional lifetime preferred for use generally in chemical kinetics and is denoted usually by $t_{½}$. It should be noted that half-life is not characteristic of a given reactant as different reactants can have the same half-life.

For first order reactions the integral rate law may be written in the following form

$$\ln\frac{[A]_0}{[A]_t} = kt \tag{2.28}$$

where $[A]_0$ and $[A]_t$ are the concentrations of a reactant A at time t = 0 and at t = t

At half-life $[A]_t = \frac{[A]_0}{2}$ and Eq. (2.28) becomes:

$$\ln 2 = kt_{1/2} \tag{2.29}$$

and

$$t_{1/2} = \frac{\ln 2}{k}$$

or

$$t_{1/2} = \frac{0.6932}{k} \tag{2.30}$$

The concentration term has therefore been eliminated on the way to half-life equation, which implies that the half-life of first order reactions is independent of the concentration of the reactant.

The half-life of second order reactions can be derived from the integrated second order rate law:

$$-\frac{1}{[A]_t} + \frac{1}{[A]_0} = -kt \tag{2.31}$$

If we apply the half-life condition that $[A]_t = \frac{[A]_0}{2}$ Eq. (2.31) yields

$$-\frac{2}{[A]_t} + \frac{1}{[A]_0} = -kt_{1/2} \tag{2.32}$$

Therefore,

$$t_{1/2} = \frac{1}{k[A]_0} \tag{2.33}$$

The equation shows that the half-life of second order increases as the initial concentration of the reactant is increased.

For zero order reactions the integrated rate law is given by:

$$[A]_t = [A]_0 - kt \tag{2.34}$$

At half-life the integrated rate equation becomes:

$$\frac{[A]_0}{2} = [A]_0 - kt \tag{2.35}$$

$$\therefore [A]_0 = 2([A]_0 - kt_{1/2}) \tag{2.36}$$

$$t_{1/2} = \frac{[A]_0}{2k} \tag{2.37}$$

From the half-life of a reaction we can make conclusion regarding the rate of the reaction. Table 2.1 shows the classification of reaction rates based on their half-lives. The table can be useful in describing photocatalytic reaction rates so that that those falling within the same category can be considered together.

Table 2.1 Classification of rates of reaction based on half-life (Wright 2006)

Half-life (s)	Time span for near-completion	Rate classification
10^{-15} to 10^{-12}	ps or less	ultra fast rate[a]
10^{-12} to 10^{-6}	μs or less	very fast rate
10^{-6} to 1	s	fast rate
1 to 10^{-3}	mins or hrs	moderate rate
10^3 to10^6	Weeks	slow rate
$> 10^6$	weeks or years	very slow rate

[a]added in this text after considering current developments in subsecond spectroscopy

2.5 Manifestations of Second Order Decay Kinetics in Photocatalysis

Second order reactions involving a single reacting entity (type I) can be represented by the chemical equation below:

$$2\text{A} \rightarrow \text{products} \tag{2.38}$$

Therefore the rate of disappearance of A is given by the equation:

$$-\frac{d[\text{A}]_t}{dt} = -k[\text{A}]^2 \tag{2.39}$$

Upon rearrangement of Eq. (2.39) we obtain

$$\frac{d[\text{A}]_t}{[\text{A}]^2} = -k dt \tag{2.40}$$

If [A] at time 0 s is $[\text{A}]_0$ and at time t seconds is $[\text{A}]_t$ we can integrate between those lower and upper limits of concentration and time

$$-\frac{1}{[\text{A}]_t} + \frac{1}{[\text{A}]_0} = -kt \tag{2.41}$$

$$\frac{1}{[\text{A}]_t} = kt + \frac{1}{[\text{A}]_0} \tag{2.42}$$

A plot of $1/[\text{A}]_t$ against time should give a straight line having intercept equal to $1/[\text{A}]_0$ and slope k. Consistence of photocatalytic degradation reactions with second order model is not expected but this unusual case was reported in the photocatalytic degradation of Auramine O aqueous solutions over ZnO catalyst owing to constraints in fitting the kinetic data to other than the second order kinetics (Kumar et al. 2007).

The recombination kinetics of charge carriers has been widely proven to be consistent with type I second order model (Serpone et al. 1995; Colombo and Bowman 1996; Bauer et al. 2004). Information gleaned from the femtosecond spectroscopic study by Colombo et al. (1995a) showed close similarities in the electron-hole kinetics and recombination dynamics of wide band gap semiconductors regardless of the photocatalyst used. In another study, Bahnemann et al. (1997) observed the recombination of trapped holes and electrons in colloidal TiO_2 within the first 200 ns. Later on, Bauer et al. (2004) studied the relaxation processes of charge carriers in 20 nm nano-crystalline ZnO at laser excitation wavelength of 750 nm. Sharp and broad peak were obtained at 385 and 600 nm respectively. The narrow peak was linked to the recombination of photo-electron and hole. The workers observed that more than 50 % of the photoactivated charge carriers disappeared in less than 10 ps. Electron and hole were rapidly trapped in about 1 ps. On the other hand, trap-assisted recombination and exciton recombination occurred within time scale of 400 and 12 ps respectively.

The femtosecond measurements of the intensity dependence of electron-hole recombination by Colombo et al. (1995b) has revealed at least two processes involved in the consumption of semiconductor-based conduction band electrons. There is a first order reaction leading to trapped electrons e^-, most likely at the surface, which subsequently undergoes very rapid second order recombination. The rate of recombination has been generally given by Eq. (2.43).

$$r_{rec} = k_{-2}[e^-][h^+] \tag{2.43}$$

Since the same number of opposite charge carriers recombine in the semiconductor particle, the rate of recombination will proceed by type I second order below.

$$r_{rec} = k_{-2}[e^-]^2 \tag{2.44}$$

Other surface reactions of second order type include the hydroxyl radical oxidation of paracetamol which has been characterised with second order rate constant of 1.7×10^9 mol^{-1} dm^{-3} s^{-1} (Yang et al. 2009). Columbo et al. (1995b) has shown that a third (or higher) order process takes place between charge carriers which leads to a long-lived species. The third-order reaction is believed to be a result of an auger process which leads to ejected electrons in deep trap states. Currently, there are no elaborate details of this third order process and to our knowledge no further investigations have been carried out.

Similarly, the mechanistic details of the second order recombination process at these microscopic scales are sketchy. However, Rothenburger et al. (1985) has monitored the dynamics of charge carrier trapping and recombination in semiconductor particle and has corroborated the charge carrier lifetime (i.e. 30 + 15 ps) and kinetic model (i.e. second order) by invoking stochastic kinetic models detailed below. In subsection 1.5.1 of Chap. 1 we have introduced transit time, which is required for charge carriers' diffusion to the surface. This parameter may vary proportionally with the radius of the semiconductor particle according to Fick's equation (Eq. 2.45) (Grätzel and Frank 1982).

$$\tau = \frac{r^2}{\pi^2 D} \tag{2.45}$$

where τ is the transit time in subseconds, r is the radius of the semiconductor particle, D is the diffusion coefficient. The recombination kinetics can be modelled by a quadratic birth and death process which implies that the exponential decay of the survival probability of a single electron-hole pair in a semiconductor particle with time. If the average lifetime of charge carrier is $\tau = k^{-1}$ and x pairs are present in a particle, the first recombination event yielding a particle with $(x-1)$ pairs will occur after an average time given by Eq. (2.46).

$$\tau = \frac{1}{x^2 k} \tag{2.46}$$

where x^2 is a factor that allows for the fact that x^2 different choices can be made from an ensemble of x electrons and x holes.

The time differential of the probability that a particle contains x electron-hole pairs ($x = 1, 2\ldots$) at time t is given by Eq. (2.47).

$$\mathrm{dP_x}(t)/\mathrm{d}t \; = k(+1)^2 \mathrm{P}_{x+1} + 1(t) - kx^2 \mathrm{P}_x(t) \tag{2.47}$$

If the charge carriers are assumed to be Poisson-distributed, this differential equation can be solved subject to the initial conditions of Eq. (2.48) for $\mathrm{P}_x(0)$.

$$\langle x \rangle(t) = \frac{\langle x \rangle_0}{1 + \langle x \rangle_0 kt} \tag{2.48}$$

The average number of charge pairs present at time t, $\langle x \rangle(t)$, can be calculated by means of the function technique to give Eq. (2.49).

$$\langle x \rangle(t) = \sum_{n=1}^{\infty} c_n e^{-n^2 kt} \tag{2.49}$$

Where

$$c_n = 2e^{-\langle x \rangle(-1)n \sum_{n=1}^{\infty} \frac{\langle x \rangle_0{}^i}{(n+i)!} \prod_{j=1}^{n} (n-i-j)}$$

The parameter $\langle x \rangle_0$ is the average number of pairs present at $t = 0$.

From the equations, two kinetic schemes may be observed. When $\langle x \rangle_0$ is very small, Eq. (2.49) becomes an exponential and the charge carrier recombination follows a first-order rate law. Secondly, the equation approximates a second-order rate model (Eq. 2.48) at high average initial occupancy of the semiconductor particles by electron-hole pairs (Fig. 2.4).

Rothenburger et al's plot of $\langle x \rangle_0 k\tau_i$ calculated from Eq. (2.49), against $\langle x \rangle_0$ for values of τ_i corresponding to first, second and third lifetimes (i.e. $i = 1/2$, 1/4 and 1/8; the fraction of initial charge carrier pairs still surviving) yielded two different limits. For first order process $\langle x \rangle_0 k\tau_i \; = \; -\ln\, i\langle x \rangle_0$, which manifested as straight lines passing through the origin. The second order recombination process was described by $\langle x \rangle_0 k\tau_i = (1/\mathrm{i}) - 1$, which appeared as horizontal lines on the plot that can approximate to second order curves.

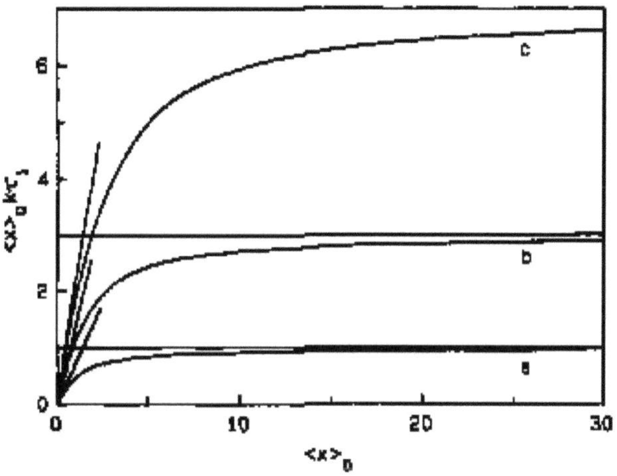

Fig. 2.4 Graph of $<x>_0 k\tau_i$ versus $<x>_0$ for values of τ_i for *a*) $i = 1/2$, *b*) $1/4$ and *c*) $1/8$. (Reproduced from Rothenberger et al. 1985)

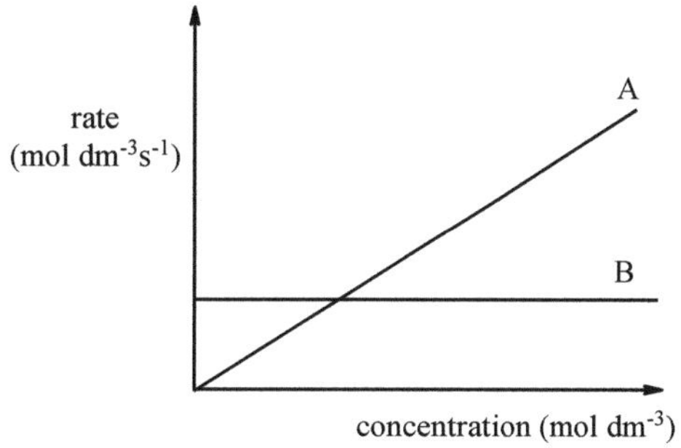

Fig. 2.5 Dependence of photocatalytic reaction rate on the concentration of reactant: first order appearance profile (*A*), zero order reaction (*B*)

2.6 Finding Order and Rate Constant From Graphic Profiles

There are several ways to analyse kinetic data in order to find out basic rate parameters. Details of methods involving calculations can be found in chemistry texts (Engel and Reid 2006; Mortimer 2008). Here, we intend to concentrate on important concepts that are critical in the studies of photocatalytic reaction rate. The methods frequently used for the determination of kinetic parameters are graphical, chosen largely because graphs allow visualisation of kinetic profile and easier quantitation of throughput of chemical reaction. For example, the graph of reaction rate versus concentration (Fig. 2.5) can help identify which kinetics plays role during chemical reaction. It can be seen from the graph the rate of zero order or pseudo zero order reactions is not affected by changes in concentration. We may wish to make comparison which curve best fits our kinetic data. If we plot kinetic data of as concentration against time for zero order and first order reactions the shape of the curves will take the form shown in Fig. 2.5.

The curve B in Fig. 2.5 displays the nature of rate versus concentration graph for zero order reactions while curve A displays the dependence of first order rate of appearance of concentration. From the figure constancy of the zero order rate as well as the independence of the rate on concentration will be readily observed. However, the curve A shows the increase in the rate of formation strictly with concentration which is consistent with the first order rate scheme.

We intend to provide the following conceptual methods of determining order and rate constant for possible use in kinetic studies of photocatalytic reactions provided that conditions of temperature, pressure and reactant doses are maintained constant.

2.6.1 Integrated Rate Law Method for Single Reactant

The integral rate law-based graphical method is one of the commonest methods available for the determination of order and rate constant. Reactions that are consistent with a given integrated rate equations will give a straight line. For the rate of disappearance the straight line will have a negative slope while for the rate of formation the graph will have a positive slope. We have earlier given a set of integrated rate laws for the zero-order reaction, first order reaction and second order reaction (Eq. 2.10, 2.15 and 2.42). The plot of ln[reactant] versus time for example gives a straight line for first order reactions and curves for zero order and second order reactions as shown in Fig. 2.6.

The inset of Fig. 2.6 shows the corresponding shape of the first order decay scheme (not drawn to scale). The slope of a tangent to the curve B at any point will give the instantaneous rate at that point. If the instantaneous rate is taken at the start of the reaction, it is referred to as the initial rate. Initial rate is identified by the steepest slope value and is frequently considered important in kinetic analyses.

2.6.2 Initial rate method

In the initial rate method the initial rate of the reaction r for the transformation of a given reactant and the [reactant] are monitored. Further, the lnr and ln[reactant] is tabulated. Now, we know that the rate of reaction r is given by

$$\mathrm{r} = k[\mathrm{reactant}]^{\mathrm{n}} \tag{2.50}$$

where n is the order and k is the rate constant of the reaction. If we take the natural logarithm of both sides of Eq. 2.50 we obtain

$$\ln \mathrm{r} = \ln k + \mathrm{n} \ln[\mathrm{reactant}] \tag{2.51}$$

By comparing Eq. (2.51) with the equation of straight line $y = mx + c$, a plot of lnr versus ln[reactant] should give a straight line with n which equals to the order of the overall reaction. The rate constant of the reaction will be obtained from the intercept.

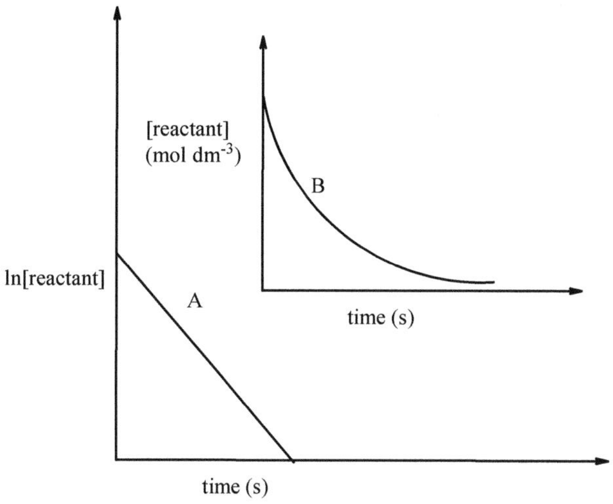

Fig. 2.6 Typical kinetic fits for first order reaction disappearance. **a** ln[reactant] against time. **b** concentration of reactant versus time

2.6.3 Initial Rate Ratio Method (For Mixed Reactants)

Supposing we are given a photocatalytic system containing relatively much lower concentration of a reactant A in presence of B, the initial rate and order with respect to the A can be obtained by isolating B (since it would almost negligibly disappear). If two different initial concentrations of A are measured, say $[A]_1$ and $[A]_2$ then the corresponding rates of the reaction would be

$$r_1 = k[A]_1^{a}[B]^{b} \tag{2.52}$$

And

$$r_2 = k[A]_2^{a}[B]^{b} \tag{2.53}$$

where a and b are reaction orders with respect to A and B, respectively.

$$\therefore \quad \frac{r_1}{r_2} = \frac{k[A]_1^{a}[B]^{b}}{k[A]_2^{a}[B]^{b}} \tag{2.54}$$

$$\Rightarrow \quad \frac{r_1}{r_2} = \frac{[A]_1^{a}}{[A]_2^{a}} \tag{2.55}$$

By taking the log of both sides of the above equation we have

$$\ln\frac{r_1}{r_2} = \ln\left(\frac{[A]_1}{[A]_2}\right)^{a} \tag{2.56}$$

Equivalently,

$$\ln\frac{r_1}{r_2} = a\ln\frac{[A]_1}{[A]_2} \tag{2.57}$$

A plot of $\ln\frac{r_1}{r_2}$ versus $\ln\frac{[A_1]}{[A_2]}$ should give a straight line through the origin with slope corresponding to the order of the reaction ‘a’.

2.7 Parametric Factors Affecting the Rate of Photocatalytic Degradation

2.7.1 *Concentration and Degradability of the Substrate*

The concentration of reactant under photocatalytic treatment can directly influence the rate of reaction by modifying photonic efficiencies (Friesen et al. 2000). At excessive substrate concentrations, the photonic efficiency would diminish and the semiconductor surface becomes saturated (Araña et al. 2004). For these reasons, photocatalytic processes offer maximum efficiency only when optimum initial concentrations are utilised. A priori, the adsorption of substrates on semiconductor photocatalysts is 'Langmuirian' and can therefore be related to the rate of reaction by Eq. (2.58).

$$r = -\frac{d[A]}{dt} = k_2 \frac{K_{ads(A)}[A]}{K_{ads(A)}[A]+1} \tag{2.58}$$

where $K_{ads(A)}$ is the adsorption constant of a species A and k_2 is the rate constant. In dilute photocatalytic solutions ($[A] < 10^{-3}$ mol dm^{-3}), $K_{ads(A)}[A]$ becomes $\ll 1$ and the reaction agrees with the pseudo-first order, while at greater concentrations (than 5×10^{-3} mol dm^{-3}), $K_{ads(A)}[A] \gg 1$ and the reaction rate scheme becomes pseudo-zero order.

Since photocatalysis is an adsorption-based process, organic molecules which can adhere effectively to the surface of the photocatalyst will be more susceptible to direct oxidation (Tariq et al. 2007). Degradability varies from one compound to another, depending upon chemical nature and adsorption behaviour. Thus, the photocatalytic degradation of aromatics depends on the substituent group. For example, it has been reported that nitrophenol is more effectively adsorbing substrate than phenol and therefore degrades faster (Bhatkhande et al. 2004). Understanding the chemistry of reactivity of organic pollutants may help in the explanation of these behaviours (Clayden et al. 2001). In the case of nitrophenol and phenol the deactivation of the phenolic ring by nitro-group may be responsible for the relatively higher degradability of the nitrophenol due to the presence of deactivating group. Similar explanation can be made on the photocatalytic degradation of other phenolics irrespective of semiconductor photocatalyst used. For instance, it has been observed that mono-chlorinated phenol regardless of substituent position degrades slower than di- or tri-chlorinated congeners (Huqul et al. 2002; Gaya 2011).

2.7.2 *Temperature*

The dependence of photocatalytic activity on temperature has been well studied and established (Kraeutler and Bard 1978; Tunesi and Anderson 1978; Fu et al. 1996; Soares et al. 2001; Evgenidou et al. 2005). The first successful equation describing the relationship between temperature and reaction rate was proposed by Arrhenius. This equation predicts an exponential dependence of reaction rate on temperature thus allowing the determination of thermodynamic properties. Arrhenius equation has been used in its common or closely related expressions as displayed below.

$$k = \mathrm{A}e^{-\frac{\mathrm{E}^*}{\mathrm{RT}}} \tag{2.59}$$

So,

$$\ln k = \ln \mathrm{A} - \frac{\mathrm{E}^*}{\mathrm{RT}} \tag{2.60}$$

where A is the frequency factor or the pre-exponential factor and E* is the activation energy. Both A and E* are characteristic of the reaction. A and E* can be obtained from the intercept and slope of a plot of ln$k$ against – 1/T, respectively. The apparent first order rate constant k should increase linearly with exp (– 1/T). Practically, E* can be calculated given rate constants k_1 and k_2 at two given temperatures T_1 and T_2 by using the Eq. (2.61)

$$\ln \frac{k_2}{k_1} = \ln \mathrm{A} - \frac{\mathrm{E}^*(\mathrm{T}_2 - \mathrm{T}_1)}{\mathrm{R}(\mathrm{T}_1\mathrm{T}_2)} \tag{2.61}$$

Photocatalytic reactions exhibit very low activation energies as they are driven by rapid electron and hole transfer processes. Apparent activation energies of photocatalytic degradation processes reported were usually small, within −8 to 11 kJ mol^{-1} below 80 °C (Bouzaida et al. 2004). However, at higher operation temperatures thermocatalysis takes the lead and activation energy can shoot up. For instance, the thermal photocatalytic oxidation of acetone on platinised titania at 100 °C was identified with activation energy up to 84 kJ mol^{-1} (San et al. 2002).

Spectral region of UV light is an important parameter of concern as this can cause rise in reaction temperature thereby increasing degradation rate. The effect of temperature on photocatalytic destruction of organics under UV-A light at ordinary to moderate temperatures is usually negligible. However, a comparison of UV-A and UV-ABC showed that the latter can be more efficient in photocatalytic organic destruction than the former alone, depending on the pollutant (Vorontsov et al. 2000). Puma and Yue (2002) observed at least two-fold increase in degradation and mineralisation rate as temperature is raised between 20 and 40 °C with concomitant use of UV-ABC light for the degradation of 2-chlorophenol. In order to comprehensively explain this observation we may liken the photocatalytic system to ordinary photochemical system such that from thermodynamic point of view the following equation can hold true:

$$\frac{\mathrm{d}\ln k}{\mathrm{dT}} = \frac{\mathrm{E}^* + \mathrm{H}^{\mathrm{o}}}{\mathrm{RT}^2} \tag{2.62}$$

where k is the photocatalytic rate constant and H^{o} is the standard enthalpy of the reaction. The equation relates thermodynamic functions with temperature gradient. The term on the left hand side of the equation is the temperature coefficient. This term becomes large positive with increase in temperature and a large positive enthalpy of photocatalytic reaction. Although photocatalytic reactions do not commonly exhibit negative temperature coefficient, Eq. (2.62) can be used to account for any such observation.

2.7.3 Light Intensity and Wavelength Effects

Heterogeneous photocatalytic processes as well as other photochemical processes are initiated by light absorption not by intermolecular collision. Consequently, photocatalytic reaction rate depends largely on the extent of radiation absorption of the photocatalyst (Curcó et al. 2002). This implies that the Grotthus–Draper law of photochemistry discussed in Chap. 1 holds for absorption of solids and the resulting photoreactions. The intensity of the irradiating light may also vary according to source and path length travelled, thus affecting the rate constant of photocatalytic reaction. This variation in intensity has consequences on the rate of photocatalytic reaction. Many workers observed enhancement in degradation rate as light intensity is increased (Karunakaran and Senthilvelan 2005; Qamar et al. 2006). Unlike in thermal catalytic reactions, the primary factor that can initiate or influence photocatalytic reactions is the light. This dependence confirms the photochemical nature of photocatalytic reactions.

Even though the nature or form of the light does not affect the reaction pathway (Stylidi et al. 2004), change in wavelength affects the initial oxidation rate and this can affect reaction pathways. In other words, the band-gap sensitization mechanism does not matter in photocatalytic degradation. But the bottom line is that only less than 5 % of the solar light can be available for UV light-responsive semiconductor photocatalyst (Karunakaran and Senthilvelan 2005). About 48 % of the solar light is in the visible range. Besides, energy loss due to light reflection, transmission and as heat is inevitable in the photoprocess (Yang and Liu 2007). This limitation largely invited more research to the topic of application of environmental semiconductor photocatalysis.

On the whole, photocatalysis is spectrum region-sensitive owing to the effect of the photocatalyst's spectral absorption. As the region of light is changed, the photocatalytic performance of semiconductor is usually dramatically affected. For instance, the photodecomposition of methyl orange on iron-doped TiO_2 nano-powder and TiO_2 that is devoid of Fe^{2+} showed dependences not only on the iron doping concentration but also on the wavelength of the irradiating light (Wang et al. 2006). Under visible light, the Fe^{2+}-doped TiO_2 had the highest photocatalytic reactivity owing to the narrowing band gap. In Chap. 5, the enhancement of photocatalysis by extending the wavelength threshold of semiconductor photocatalysts is discussed in details.

The Stark–Einteins law of photochemical equivalence restricts one light quantum to a molecule. Practically, the statement of the law requires adjustment to suit most systems including semiconductor photocatalytic systems. In heterogeneous photocatalysis, electron transfer processes may be hampered by recombination of charge carriers and therefore effective photoactivation may not simply require one light quantum. This fact led to the necessity of the measurement of intensity of light and the estimation of the efficiency of photocatalytic processes.

2.7.3.1 Photonic Efficiencies

Basically, the overall quanta of light absorbed by photochemical systems can be represented by quantum yield Φ, a ratio expressed as follows:

$$\Phi = \frac{\text{rate of photo-induced reaction}}{\text{flux of absorbed radiation}} \tag{2.63}$$

Therefore, for the degradation of a compound A, over irradiation time t with photon intensity I, Φ can mathematically be expressed by Eq. (2.64). The Φ value of ideal efficient system is unity. However, majority of the known photocatalytic systems do not exhibit such a quantum yield.

$$\Phi = \left(\frac{d[A]}{dt}\right) I^{-1} \tag{2.64}$$

The dose of light absorbed by a photocatalytic system can only be determined if the intensity of the incident or absorbed photon is known. These variables can be measured using a probe known as the actinometer (also called dosimeter). An actinometer can be defined as either a physical device or a chemical system that allows the measurement of incident photons or photons absorbed in a defined space per unit time. This definition distinguishes two actinometers: physical actinometers and chemical actinometers. Irrespective of type, all actinometers meant largely for photochemical studies are only applicable in the UV to visible spectral region. The physical actinometers used in radiant energy measurements come in many different forms such as bolometer, photodiode, pyranometer and quantum counter. These devices have the ability to show signals equivalent to the incident photon flux. For example, quantum counter uses rhodamine B to absorb incident photons and emit fluorescence which would finally be detected by a photomultiplier. Many workers however prefer to use chemical actinometers for the measurement of radiant energy as these systems measure directly the absorbed light flux.

Chemical actinometry is usually based on a well characterised photosensitive gas mixtures or solutions. The photosensitivity of such a reaction depends upon the photochemical nature of the chemical system that forms the actinometer. Basically, photosensitivity of chemical dosimeters occurs at a given wavelength range and associated with known quantum yield value. Chemical systems used in actinometry come with different reactions such as photoisomerisation, photolysis, photoionisation, photoreversible reaction and photochromism. The experimental procedures and methods of calculation of the radiation intensity for several actinometers have been well outlined by Kuhn et al. (2004). Here we highlight some fundamental principles of selected chemical actinometric systems the reader must know.

An actinometer with as many desirable features as possible will be preferred by most researchers. One important feature that is often considered is the wavelength range. Table 2.2 displays with properties together, the commonest actinometers used in photochemical laboratories.

The uranyl oxalate or dioxouranium(VI) oxalate actinometer consists of 0.05 mol dm^{-3} oxalic acid photolysed in the presence of 0.01 mol dm^{-3} uranyl sulphate. The decomposition of oxalic acid yields carbon oxides as shown in the following chemical equation:

$$H_2C_2O_4 + \xrightarrow{hv/UO_2^{2+}} H_2O + CO_2 + CO \qquad (2.65)$$

The uranyl sulphate acts as sensitiser but the mechanism for the sensitisation is not exactly known. There is claim that the UO_2^{++} forms complex with the oxalic acid that breaks down to yield H_2O, CO_2, CO as products.

$$[UO_2(C_2O_4)_2]^{2-} \rightarrow H_2O + CO_2 + CO \qquad (2.66)$$

The extent of the decomposition of oxalic acid in the process can be determined by the back titration of the residual $H_2C_2O_4$ with $KMnO_4$ solution (a relatively insensitive method) or $CeSO_4$ (a method with poor accuracy), or reading the absorbance (at 320 nm) of the solutions previously photolysed and incubated with $CeSO_4$ (in H_2SO_4) at low temperatures for a few minutes. This actinometer has the advantage of being a wide-range actinometer, covering wavelengths from 200 to 500 nm.

$$2Ce^{4+} + C_2O_4{}^{2-} \rightarrow 2Ce^{3+} + 2CO_2 \qquad (2.67)$$

Perhaps the most popular photolysis-based actinometer is the potassium ferrioxalate, first reported by Hatchard and Parker (1956). Hatchard–Parker actinometer is based on the well-studied photolysis of $K_3[Fe(C_2O_4)_3].3H_2O$ which the pioneering investigators proposed to proceed as follows:

$$[Fe(C_2O_4)_3]^{3-} \xrightarrow{hv} Fe^{2+} + 2C_2O_4{}^{2-} + C_2O_4{}^{\bullet-} \qquad (2.68)$$

$$[Fe(C_2O_4)_3]^{2-} + C_2O_4{}^{\bullet-} \rightarrow Fe^{2+} + 3C_2O_4{}^{-} + CO_2 \qquad (2.69)$$

$$C_2O_4{}^{\bullet-} + O_2 \rightarrow O_2{}^{\bullet-} + 2CO_2 \qquad (2.70)$$

Equilibrium between mono, bis, and tris oxalato complexes present at the pH of the actinometry solutions takes place according to the following equations:

$$[Fe(C_2O_4)_3]^{3-} \xrightarrow{hv} [Fe(C_2O_4)_2]^{2-} + C_2O_4{}^{-} \qquad (2.71)$$

$$[Fe(C_2O_4)_3]^{3-} + C_2O_4{}^{-} \rightarrow [Fe(C_2O_4)_3]^{2-} + C_2O_4{}^{2-} \qquad (2.72)$$

$$[Fe(C_2O_4)_3]^{2-} \rightarrow [Fe(C_2O_4)_2]^{2-} + 2CO_2 \qquad (2.73)$$

A complex of 1,10-phenanthroline is formed with Fe^{2+} which (the complex) absorbs at 510 nm. However, as $K_3[Fe(C_2O_4)_3]$ and 1,10-phenanthroline solutions are

Table 2.2 Properties of some popular wide-range chemical actinometers used in photocatalysis

Actinometer	Mechanism of action	Φ and wavelength range	Remark
Ferrioxalate actinometer	Photolysis of $K_3[Fe(C_2O_4)_3]$	$\Phi = 1.25$ to 0.9 $\lambda = 250$ to 465 nm	Covers UV and visible region. Can work at wavelengths up to 577 nm with less accuracy (owing to smaller molar absorptivities) (Kuhn et al. 2004) or down to 220 nm with procedure modification (Fernandez et al. 1979)
Uranyl oxalate	Photolysis of $H_2C_2O_4$	$\Phi = 0.5 + 2$ $\lambda = 200$ to 500	A UV actinometer. About half as sensitive as ferrioxalate actinometer and $CeSO_4$ is usually impure (Kuhn et al. 2004). Uranium hazard
Reinecke's salt	photoaquation reaction of Reinecke's complex $K[Cr(NH_3)_2(SCN)_4]$ upon irradiation which releases SCN-ion	315 to 600 nm	For use in visible region. Procedure not simple as SCN^- may be formed by fast thermal reactions (Santos et al. 2009)
Aberchrome actinometer		$\Phi_{310-370} = 0.2$ $\lambda = 310$ to 370 nm	Photodegradation occurs below 300 nm. Actinometer can only be used once in the UV range since the *E* isomer forms Z isomers which both do not absorb in the visible range.[a] $\Phi_{365.6} = 0.13$ ($E \rightarrow Z$), $\Phi_{365.6} = 0.12$ ($Z \rightarrow E$) (Montalti et al. 2006)

Table 2.2 (continued)

Actinometer	Mechanism of action	Φ and wavelength range	Remark
Iodide-Iodate actinometer	Based on yields of I_3^- from KI (0.6 mol dm^{-3})/ KIO_3(0.1 mol dm^{-3}) system in 0.01 mol dm^{-3} $Na_2B_4O_7 \cdot 12\,H_2O$ buffer at pH 9.25. Absorbance of I_3^- and I^- can be read spectrophotometrically at 352 and 300 nm respectively	$\Phi = 0.92$ $\lambda = 205$ to 245 nm	Short range UV actinometer. (Goldstein and Rabbani 2008)

[a](*E*) stands for "*Entgegen*", a German name for *cis*, while (*Z*) stand for "*Zuzammen*" meaning *trans* isomer

extremely that photosensitive that they must be prepared fresh before actinometric experiments. Ferrioxalate actinometer has many advantages which include being thermally stable, insensitive to the order of reagent addition, easily reproducible and responsive over a wide range of wavelength. However, the use of actinometer entails the addition of several chemical agents, dark experimental conditions or requirement of red safe light, long equilibration time (at least half an hour, to confirm the complete colourisation of Fe^{2+}-phenanthroline complex), which are laborious, time consuming and possible sources of experimental errors (Lee et al. 2007).

Fulgide or (*E*)-2-[1-(2,5-dimethyl-3-furylethylidene]-3-isopropylidene succinic anhydride is a photochromic compound that used to appear in the market with trade name Aberchrome 540. Photochromism is the reversible transformation of compound between two molecular forms having different absorption spectra, induced in one or both directions by the absorption of light. In open form fulgide is pale yellow and can undergo photocyclisation reaction when irradiated with UV light of wavelength 310 to 370 nm to yield thermally stable closed red form 7,7a-dihydro-2,4,7,7,7a-pentamethylbenzo(b)furan-5,6-dicarboxylic anhydride. On irradiation with visible light of wavelength 435 to 546 nm the reaction is reversed.

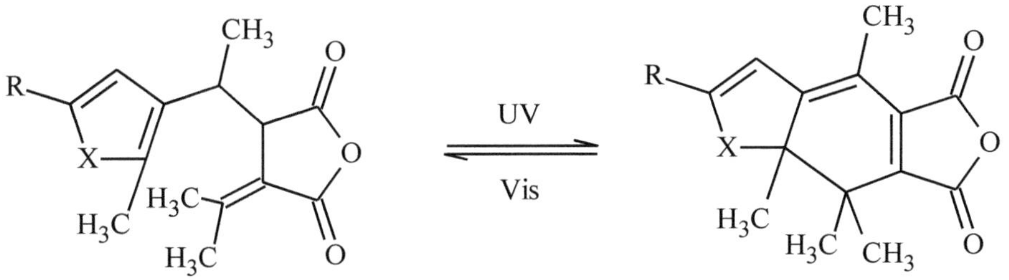

(2.74)

As Aberchrome actinometer works in the UV-A region, it has been used by many workers to determine the amount of incident radiation that can be absorbed by the photocatalyst in the same region (Muneer et al. 2001).

Azobenzene actinometer as the name implies is based on the photoreversible isomerisation reaction of azobenzene. The actinometer works within 230 to 460 nm. This actinometer has the following properties (Kuhn et al. 2004). In methanol or isooctane the actinometer absorbs at 358 nm and the trans→cis change proceeds with $\Phi = 0.1 - 0.2$ while the reverse cis→trans reaction proceeds with $\Phi = 0.4 - 0.5$. Many organic solvents such as n-hexane, cyclohexane, dilute ethanol (20 %), tetrahydrofuran (THF) and acetonitrile can be used as solvents.

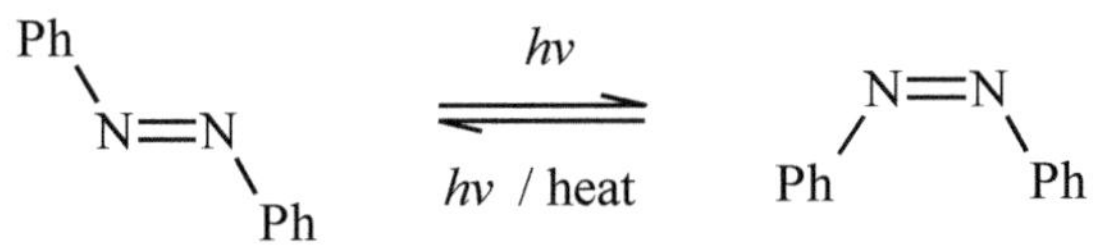

(2.75)

The possibility of exploitation of nitrate photolysis in chemical actinometry has been demonstrated by Jankowski et al. (1999). About 90 % of nitrate photolyses according to the following chemical equations:

$$NO_3^- + h\nu \rightarrow NO_2 + O^{\bullet -} \tag{2.76}$$

Similarly, nitrite may also undergo direct photolysis according to Eq. (2.77)

$$NO_2^- + h\nu \rightarrow NO + O^{\bullet -} \tag{2.77}$$

The resulting $O^{\bullet -}$ radical readily gets protonated to yield $^{\bullet}OH$ which combined with benzoic acid provides a satisfactory actinometer. The hydroxyl radical can be scavenged by benzoic acid to yield highly fluorescent and chromaphoric *o*-hydroxybenzoic acid (or salicylic acid) and *p*-hydroxybenzoic acid as shown in Eq. (2.78).

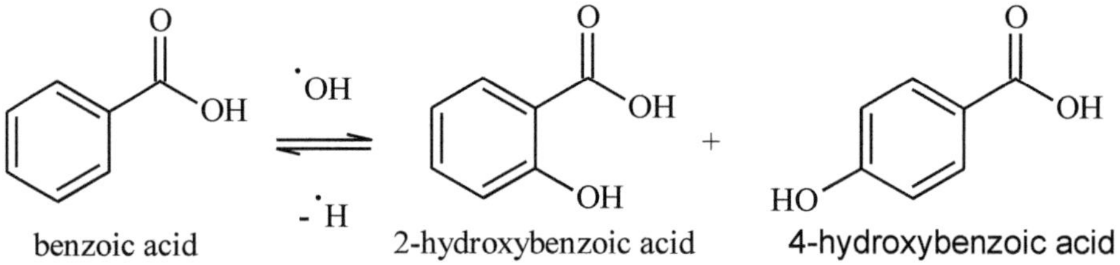

(2.78)

Quantum yields (at 25 °C) for salicylic acid production from both nitrate and nitrite photolysis are low (0.0014 between 290 and 330 nm; and 0.024 to 0.078 between 298 and 390 nm, respectively) (Jankowski et al. 1999). In addition, the quantum yields for salicylic acid production were approximately 50–60 % greater than quantum yields for *p*-hydroxybenzoic acid production from nitrate and nitrite photolysis. Because of low quantum yields, these actinometers are insensitive to laboratory illumination and can maintain initial rate conditions for hours. However, one disadvantage that must be highlighted is that both nitrate and nitrite actinometers show thermal dependence of quantum yield.

In actual fact, the photon flux absorbed by the photocatalytic reaction may be less than that estimated by actinometry owing to the presence of semiconductor particles (which may cause dispersion of light) and the absorption of photons by the walls of the photoreactor or lamp. Generally in semiconductor-based heterogeneous systems, quantum yield is itself experimentally difficult to determine exactly due to light scattering and thermal electron-hole recombination among other things (Serpone et al. 1996; Aguado et al. 2006; Silva et al. 2006; Minero and Vione 2006). As a result, relative photonic efficiency ξ_r has been proposed which is experimentally measurable and can be related to the overall quantum yield by the following equation (Salinaro et al. 1999):

$$\Phi = \xi_r \Phi_{phenol} \tag{2.79}$$

where Φ_{phenol} is the quantum yield for the photocatalysed oxidative disappearance of phenol over Degussa P-25 TiO_2.

In the measurement of photonic parameters phenol and Degussa P-25 TiO_2 are employed to serve as primary standard actinometric molecule and catalyst material, respectively. Some workers still prefer to use photonic efficiency because both the photonic efficiency and the relative photonic efficiency have one or two disadvantages. The latter has the advantage of being reactor-independent. Moreover, it allows comparison of process efficiencies for the photodegradation of aromatic substrates in different heterogeneous mediums such as dispersions, immobilized photocatalyst systems and solid/gas or solid/solution (Serpone and Salinaro 1999). Thus, relative photonic efficiency permits comparison of different photocatalytic experiments. However, for comparing experiments under fluctuating radiation relative photonic efficiency cannot be used straightaway. At any rate, photonic efficiency defined as the photocatalytic degradation rate per incident light intensity, has been thus far widely used in many laboratories (Serpone et al. 1996; Muneer et al. 2001; Serpone 1997; Tahiri et al. 1996). This parameter differs from and must not be confused with apparent quantum yield or photonic efficiency, defined as the moles of molecule consumed per amount of incident photons in Einstein, whose usage has been discontinued (Serpone and Salinaro 1999).

In order to derive synergy, photocatalysis may be coupled with certain physical operations. This practice makes it difficult to have a single equation to represent quantum efficiency. In photoelectrocatalysis and photovoltaics, an alternative conversion efficiency is used which is based on incident photon to current efficiency (IPCE). The incident photon to current efficiency gives the overall quantum yield of the system. This parameter is encapsulated by Eq. (2.80) (Beranek et al. 2007).

$$\text{IPCE} = \frac{i_{ph} h\nu}{I_0 \lambda} \tag{2.80}$$

where $h\nu$ can be given the value of 1240 eV (Vinodgopal et al 1996). The incident photon to current efficiency is equivalently called external quantum efficiency (EQE). Leng et al. (2003) have related quantum yield to IPCE by Eq. (2.81).

$$\Phi = \phi_{\text{eff}} \times \text{IPCE} \tag{2.81}$$

where ϕ_{eff} is the photocurrent efficiency and IPCE is the ratio of incident photon i_{ph} to incident light intensity I_0. The photocurrent efficiency ϕ_{eff} is related to the concentration of pollutant by Eq. (2.82).

$$\phi_{\text{eff}} = \frac{\Delta C V F}{Q_t} \tag{2.82}$$

where C is the concentration of substrate, V is the solution volume, F is Faraday's constant (96,500 C mol^{-1}). The quantity of current passed Q_t is given in terms of the incident photon i_{ph}, the time difference from the initial time by Eq. (2.83).

$$Q_t = i_{\text{ph}}(t - t_0) \tag{2.83}$$

2.7.4 Effect of Oxygen

Oxygen is the safest, cheapest and ready available electron acceptor for use in photocatalytic degradation. Oxygen can influence the rate of photocatalytic reaction positively by scavenging electron to form superoxide. Oxygen reduction reaction or more comprehensively, the adsorption of scavengers on semiconductor surfaces has been shown to be the rate-controlling process in photocatalytic decomposition (Wang et al. 1992; Kormann et al. 1991). Apart from scavenging function, the presence of oxygen is critical to the degradation of certain intermediate products. Studies have confirmed that there is no difference in the rate was observed when either natural air or O_2 sparging was used in the reactivity experiment provided the delivery is constant and sufficient (McMurray et al. 2006). In general, if dioxygen is constantly supplied, the coverage of semiconductor surface by adsorbed oxygen θO_2, is also assumed to be constant and this can be included in the rate law (Herrmann 1999). For substrate (A), decomposed under constant oxygen supply according to Eq. (2.84) the initial rate r is given by Eq. (2.85).

$$A + O_2 \rightarrow \text{products} \tag{2.84}$$

$$r = \frac{-d[A]}{dt} = k\theta O_2 \theta_A = k'\theta_A \tag{2.85}$$

where k' is the apparent rate constant and θ_A is the fractional surface coverage of semiconductor by substrate. k' depends upon light intensity and oxygen coverage. Values of the rate constant k' were reported by several workers to be in the order of 10^3 mol dm^{-3} (Hoffmann et al. 1995). Where there is variation in oxygen partial pressure, the kinetic effect of oxygen as well as values of θO_2 can be given by a simple Langmuir adsorption isotherm represented by Eq. (2.86).

$$\theta_{O_2} = \frac{K_{ads(O_2)}[O_2]}{1 + K_{ads(O_2)}[O_2]} \tag{2.86}$$

If the activity of oxygen is measured as partial pressure, then θ_{O2} can be expressed as follows.

$$\theta_{O_2} = \frac{K_{ads(O_2)} P_{O_2}}{1 + K_{ads(O_2)} P_{O_2}} \tag{2.87}$$

By plugging Eq. (2.87) in Eq. (2.85) we arrive at the equation relating photocatalytic rate with oxygen partial pressure:

$$\frac{1}{r} = \frac{1}{k' K_{ads(O_2)} P_{O_2} \theta_A} + \frac{1}{k\theta_A} \tag{2.88}$$

The reciprocal of initial rate can be plotted against that of the partial pressure to obtain the values of the constants in the equation from slope and intercept.

2.7.5 Concentration of Photocatalyst and Surface Properties

A very important parameter that influences the performance of semiconductor photocatalyst in photocatalytic degradation is the surface morphology. Surface morphology is a property that is dictated by particle size and agglomerate size (Dinga et al. 2005). Consequently, modification of surface morphology of semiconductor photocatalysts can give rise to a more efficient form for photocatalytic application. For example, Li et al. (2005) observed a change in the prismatic morphology of ZnO particles to sheet shape during coupling with vanadium oxide, and to rod-like shape morphology during coupling with cesium oxide. The ZnO-vanadium oxide composite resulted in higher photocatalytic activity for acetaldehyde degradation.

The number of photons absorbed by the same semiconductor photocatalyst depends partly upon particles size. For tiny semiconductor particles and assuming negligible light scattering the number of photons absorbed by a photocatalyst particle per second can be expressed as a function of the semiconductor particle size and light intensity by Eq. 2.89 (Klabunde and Richards 2009).

$$\nu = d^3\alpha(\lambda)/\ln 10 \tag{2.89}$$

where d is the particle size of the semiconductor cross-section, I is the light intensity and α is the coefficient of light transmission at fixed wavelength. However, no direct relationship between photocatalytic activity and crystallinity or surface area has been established for differently prepared semiconductor powders even with the same method but with particle morphology (Li and Haneda 2003).

The rate of photocatalytic reaction is strongly influenced by concentration of the photocatalyst. In heterogeneous photocatalytic systems, the optimum catalyst concentration must be determined, in order to avoid light scattering and to have effective concentration of uncovered catalyst surface. Unless that is achieved there will be poor absorption of photons overall, leading to decrease in degradation rate (Chun et al. 2000; Saquib and Muneer 2003; Dodd et al. 2005). As sufficient catalyst loading is reached, agglomerates are believed to be formed whose size R, can be estimated using Eq. 2.90 (Mehrotra et al. 2003).

$$R = \frac{3}{\eta}\sqrt{\frac{D}{k\rho_{sc}A_s}} \tag{2.90}$$

where η is the optimal catalyst domain, k is the degradation rate constant, S_a (m^2/g) is the specific surface area of the reactor, D is the diffusion coefficient and ρ_{sc} is the density of the photocatalyst. The excessive agglomeration of semiconductor catalysts is a negative development that will translate to reduction in photocatalytic rate.

2.8 Conclusion

In heterogeneous systems, several factors may directly or indirectly affect the rate of reaction. The amount of photocatalyst, pH and the concentration of the compound targeted for destruction are the major factors commonly studied. The relationship between efficiency and these factors has been classically established by the 'one factor at a time' (OFAT) method. The methods based on OFAT are time consuming and may entail the waste of chemicals. If not viewed with the right lenses, these methods may also lead to misinterpretation of results. Recently, experimental design has attracted much interest for use in optimisation of photocatalytic processes. Response surface and models based on actual experiments have shown promising applicability in this context. Probably in the nearer future, statistical approaches may reign as methods of optimisation of parameters in both synthesis of the photocatalysts and the photocatalytic process.

References

Aguado J, Grieken R van, López-Munõz M-J et al (2006) A comprehensive study of the synthesis, characterization and activity of TiO_2 and mixed TiO_2/SiO_2 photocatalysts. Appl Catal A 312:202–212

Araña J, Nieto JLM, Melián JAH (2004) Photocatalytic degradation of formaldehyde containing wastewater from veterinarian laboratories. Chemosphere 55:893–904

Augugliaro V, Kisch H, Loddo V et al (2008) Photocatalytic oxidation of aromatic alcohols to aldehydes in aqueous suspension of home-prepared titanium dioxide 1. Selectivity enhancement by aliphatic alcohols. Appl Catal A 349:182–188

Bahnemann DW, Hilgendorff M, Memming R (1997) Charge carrier dynamics at TiO_2 particles: reactivity of free and trapped holes. J Phys Chem B 101:4265–4275

Bauer C, Boschloo G, Mukhtar E et al (2004) Ultrafast relaxation dynamics of charge carriers relaxation in ZnO nanocrystalline thin films. Chem Phys Lett 387:176–181

Bayarri B, Abellán MN, Giménez J et al (2007) Study of the wavelength effect in the photolysis and heterogeneous photocatalysis. Catal Today 129:231–239

Beranek R, Neumann B, Sakthivel S et al (2007) Exploring the electronic structure of nitrogen-modified TiO2 photocatalysts through photocurrent and surface photovoltage studies. Chem Phys 339:11–19

Bhatkhande DS, Kamble SP, Sawant SB et al (2004) Photocatalytic and photochemical degradation of nitrobenzene using artificial ultraviolet light. Chem Eng J 102:283–290

Bouzaida I, Ferronato C, Chovelon JM et al (2004) Heterogeneous photocatalytic degradation of the anthraquinonic dye, Acid Blue 25 (AB25): a kinetic approach. J Photochem Photobiol A 168:23–30

Chun H, Yizhong W, Hongxiao T (2000) Destruction of phenol aqueous solution by photocatalysis or direct photolysis. Chemosphere 41:1205–1209

Clayden J, Warren S, Greeves N et al (2001) Organic chemistry. Oxford University Press, New York

Colombo DP Jr, Bowman RM (1996) Does interfacial charge transfer compete with charge carrier recombination? A femtosecond diffuse reflectance investigation of TiO_2 nanoparticles. J Phys Chem 100:18445–18449

Colombo DP Jr, Skinner DE, Cavaleri JJ et al (1995a) Femtosecond spectroscopy of quantum- and bulk-sized semiconductor particles. Paper presented at the 210th ACS National Meeting, American Chemical Society. 20–24 August, Chicago, 1995

Colombo DP Jr, Roussel KA, Saeh J et al (1995b) Femtosecond study of the intensity dependence of electron-hole dynamics in TiO_2 nanoclusters. Chem Phys Lett 232:207–214

Curcó D, Giménez J, Addardak A (2002) Effects of radiation absorption and catalyst concentration on the photocatalytic degradation of pollutants. Catal Today 76:177–188

Dinga H, Suna H, Shan Y (2005) Preparation and characterization of mesoporous SBA-15 supported dye-sensitized TiO_2 photocatalyst. J Photochem Photobiol A 169:101–107

Dodd AC, McKinley AJ, Saunders M et al (2006) Effect of particle size on the photocatalytic activity of nanoparticulate zinc oxide. J Nanopart Res 8:43–51

Engel T, Reid P (2006) Physical chemistry. Pearson Education, San Fransisco

Evgenidou E, Fytianos K, Poulios I (2005) Semiconductor-sensitized photodegradation of dichlorvos in water using TiO_2 and ZnO as catalysts. Appl Catal B 59:81–89

Fernández E, Figuera JM, Tobar A (1979) Use of potassium ferrioxalate actinometer below 254 nm. J Photochem 11:69–71

Friesen DA, Morello L, Headleya JV et al (2000) Factors influencing relative efficiency in photo-oxidations of organic molecules by $Cs_3PW_{12}O_{40}$ and TiO_2 colloidal photocatalysts. J Photochem Photobiol A 133:213–220

Fu X, Clark LA, Zeltner WA et al (1996) Effects of reaction temperature and water vapor content on the heterogeneous photocatalytic oxidation of ethylene. J Photochem Photobiol A 97:181–186

Gaya UI (2011) Comparative analysis of ZnO−catalyzed photo−oxidation of *p*−chlorophenols. Eur J Chem 2:163–167

Goldstein S, Rabani J (2008) The ferrioxalate and iodide–iodate actinometers in the UV region. J Photochem Photobiol A 193:50–55

Grätzel M, Frank AJ (1982) Interfacial electron-transfer reactions in colloldal semiconductor dispersions. Kinetic Analysis. J Phys Chem 86:2964–2967

Hatchard CG, Parker CA (1956) A newsensitive chemical actinometer. II. Potassium, ferric oxalate as a standard chemical actinometer. Proc R Soc London Ser A 235:518–536

Herrmann J-M (1999) Heterogeneous photocatalysis: fundamentals and applications to the removal of various types of aqueous pollutants. Catal Today 53:115–129

Hoffmann MR, Martin ST, Choi W et al (1995) Environmental applications of semiconductor photocatalysis. Chem Rev 95:69–96

Huqul M, Ercaq E, Apak R (2002) Kinetic studies on UV-photodegradation of some chlorophenols using TiO_2 catalyst. J Environ Sci A: Tox Hazard Subst Environ Eng 37(3):365–83

Jankowski JJ, Kieber DJ, Mopper K (1999) Nitrate and nitrite ultraviolet actinometers. Photochem Photobiol 70:319–328

Karunakaran C, Senthilvelan S (2005) Photooxidation of aniline on alumina with sunlight and artificial UV light. Catal Commun 6:159–165

Klabunde KJ, Richards RM (eds) (2009) Nanoscale materials in chemistry. Wiley, New Jersey

Kormann C, Bahnemann DW, Hoffman MR (1991) Photolysis of chloroform and other organic molecules in aqueous TiO_2 suspensions. Environ Sci Technol 25:494–500

Kraeutler B, Bard AJ (1978) Heterogeneous photocatalytic decomposition of saturated carboxylic acids on TiO_2 powder decarboxylative route to alkanes. J Am Chem Soc 100(19):5958

Kuhn HJ, Braslavsky SE, Schmidt R (2004) Chemical actinometry (IUPAC technical report). Organic and biomolecular chemistry division, subcommittee on photochemistry, International Union of Pure and Applied Chemistry. Pure Appl Chem 76:2105–2146

Kumar KV, Porkodi K, Selvaganapathi A (2007) Constrain in solving Langmuir-Hinshelwood kinetic expression for the photocatalytic degradation of Auramine O aqueous solutions by ZnO catalyst. Dyes Pigm 75:246–249

Lee J, Kim J, Choi W (2007) Ferrioxalate-polyoxometalate system as a new chemical actinometer. Environ Sci Technol 41:5433–5438

Leng WH, Zhang Z, Zhang JQ (2003) Photoelectrocatalytic degradation of aniline over rutile TiO_2/Ti electrode thermally formed at 600°C. J Mol Catal A: Chem 206:239–252

Li D, Haneda H (2003) Morphologies of zinc oxide particles and their effects on photocatalysis. Chemosphere 51:129–137

Li D, Haneda H, Ohashi N et al (2005) Morphological reform of ZnO particles induced by coupling with MO_x (M = V, W, Ce) and the effects on photocatalytic activity. Thin Solid Films 486:20–23

McMurray TA, Dunlop PSM, Byrne JA (2006) The photocatalytic degradation of atrazine on nanoparticulate TiO_2 films, J Photochem Photobiol A 182:43–51

Mehrotra K, Yablonsky GS, Ray AK (2003) Kinetic studies of photocatalytic degradation in a TiO_2 Slurry System: Distinguishing working regimes and determining rate dependences. Ind Eng Chem Res 42:2273–2281

Minero C, Vione D (2006) A quantitative evaluation of the photocatalytic performance of TiO_2 slurries. Appl Catal B 67:257–269

Montalti M, Credi A, Prodi L et al (2006) Handbook of Photochemistry 3rd edn. Taylor and Francis, USA

Mortimer RG (ed) (2008) Physical chemistry. Elsevier Academic Press, Canada

Muneer M, Theurich J, Bahnemann D (2001) Titanium dioxide mediated photocatalytic degradation of 1,2-diethyl phthalate. J Photochem Photobiol A 143:213–219

Muradov NZ, Raissi AT, Muzzey D et al (1996) Selective photocatalytic destruction of VOCs. Sol Energy 56(5):445–453

Puma GL, Yue PL (2002) Effect of the radiation wavelength on the rate of photocatalytic oxidation of organic pollutants. Ind Eng Chem Res 41:5594–5600

Qamar M, Muneer M, Bahnemann D (2006) Heterogeneous photocatalysed degradation of two selected pesticide derivatives, triclopyr and daminozide in aqueous suspensions of titanium dioxide. J Environ Manage 80:99–106

Rothenberger G, Moser J, Gratzel M et al (1985) Charge carrier trapping and recombination dynamics in small semiconductor particles. J Am Chem Soc 107:8054–8059

Salinaro A, Emeline AV, Zhao J et al (1999) Terminology, relative photonic efficiencies and quantum yields in heterogeneous photocatalysis. Part I: experimental determination of quantum yield. Pure Appl Chem 71:321–335

San N, Hatipoğlu A, Koçtürk G et al (2002) Photocatalytic degradation of 4-nitrophenol in aqueous TiO_2 suspensions: theoretical prediction of the intermediates. J Photochem Photobiol A 146:189–197

Santos AR, Ballardini R, Belser P et al (2009) Photochemical investigation of a photochromic diarylethene compound that can be used as a wide range actinometer. Photochem Photobiol Sci 8:1734–1742

Saquib M, Muneer M (2003) TiO_2-mediated photocatalytic degradation of a triphenyl methane dye (gentian violet), in aqueous suspensions. Dyes Pigm 56:37–49

Serpone N (1997) Relative photonic efficiencies and quantum yields in heterogeneous photocatalysis. J Photochem Photobiol A 104:1–12

Serpone N, Lawless D, Khairutdinov R et al (1995) Subnanosecond relaxation dynamics in TiO_2 colloidal sols (Particle sizes R, = 1.0-13.4 nm). Relevance to heterogeneous photocatalysis. J Phys Chem 99:16655–16661

Serpone N, Salinaro A (1999) Terminology, relative photonic efficiencies and quantum yields in heterogeneous photocatalysis. Part I: suggested protocol (Technical Report). Pure Appl Chem 71:303–320

Serpone N, Sauve G, Koch R et al (1996) Standardization protocol of process efficiencies and activation parameters in heterogeneous photocatalysis: relative photonic efficiencies ζr. J Photochem Photobiol A 94:191–203

Silva CG, Wang W, Faria JL (2006) Photocatalytic and photochemical degradation of mono-, di- and tri-azo dyes in aqueous solution under UV irradiation. J Photochem Photobiol A 181:314–324

Soares ET, Lansarin MA, Moro CC (2001) A study of process variables for the photocatalytic degradation of Rhodamine B. Braz J Chem Eng 24(01):29–36

Stylidi M, Kondarides DI, Verykios XE (2004) Visible light-induced photocatalytic degradation of Acid Orange 7 in aqueous TiO_2 suspensions. Appl Catal B 47:189–201

Tahiri H, Serpone N, Le Mao R (1996) Application of concept of relative photonic efficiencies and characterization of a new titania photocatalyst designed for environmental remediation. J Photochem Photobiol A 93:199–203

Tariq MA, Faisal M, Muneer M et al (2007) Photochemical reactions of a few selected pesticide derivatives and other priority organic pollutants in aqueous suspensions of titanium dioxide. J Mol Catal A: Chem 265:231–236

Theurich J, Lindner M, Bahnemann DW (1996) Photocatalytic degradation of 4-chlorophenol in aerated aqueous titanium dioxide suspensions: a kinetic and mechanistic study. Langmuir 12:6368–6376

Tunesi S, Anderson MA (1978) Photocatalysis of 3,4-DCB in TiO_2 aqueous suspension; Effects of temperature and light intensity; CIR-FTIR interfacial analysis. Chemosphere 16(7):1447–1456

Vinodgopal K, Bedja I, Kamat PV (1996) Nanostructured semiconductor films for photocatalysis. Photoelectrochemical behavior of SnO_2/TiO_2 composite systems and its role in photocatalytic degradation of a textile azo dye. Chem Mater 8:2180–2187

Vorontsov AV, Stoyanova IV, Kozlov DV et al (2000) Kinetics of the photocatalytic oxidation of gaseous acetone over platinized titanium dioxide. J Catal 189:360–369

Wang C-M, Heller A, Gerischer H (1992) Palladium catalysis of O_2 reduction by electrons accumulated on TiO_2 particles during photo-assisted oxidation of organic compounds. J Am Chem Soc 114:5230–5234

Wang XH, Li JG, Kamiyama H et al (2006) Wavelength-Sensitive Photocatalytic degradation of methyl orange in aqueous suspension over Iron(III)-doped TiO_2 Nanopowders under UV and visible light irradiation. J Phys Chem B 110:6804–6809

Wilke K, Breuer HD (1999) The influence of transition metal doping on the physical and photocatalytic properties of titania. J Photochem Photobiol A 121:49–53

Wright MR (2006) An Introduction to chemical kinetics. Wiley, England

Yang L, Liu Z (2007) Study on light intensity in the process of photocatalytic degradation of indoor gaseous formaldehyde for saving energy. Energ Convers Manage 48:882–889

Yang L, Yu LE, Ray MB (2009) Photocatalytic oxidation of paracetamol: dominant reactants, intermediates, and reaction mechanisms. Environ Sci Technol 43:460–465

Yang X, Tamai N (2001) How fast is interfacial hole transfer? In situ monitoring of carrier dynamics in anatase TiO_2 nanoparticles by femtosecond laser spectroscopy. Phys Chem Chem Phys 3:3393–3398

Chapter 3
Mechanistic Principles of Photocatalytic Reaction

Abstract As developments in semiconductor photocatalysis flourish, profound success is being recorded in the elucidation of reactivity vis-à-vis mechanism. The mechanism of photocatalytic reactions lies between two limits. From viewpoint of photochemical kinetics, the reaction mechanism will be controlled by the properties of light and consequences of light absorption. However as mass transfer dominates, the process will be controlled by the non-catalytic diffusion mechanism. This chapter highlights both diffusion and photocatalytically controlled kinetics in semiconductor-based heterogeneous regimes. In the final place, some conventional analytical methods for validation of photocatalytic reaction mechanism are discussed.

3.1 Background Sketch

Reaction mechanism provides the details of steps involved in overall chemical reaction. It highlights both the reaction progress and the substances formed en route to the formation of products. These information directly affect the understanding of the process. There are three important theories of reaction rate which need to be briefly introduced: the transition state theory, the collision theory and the Arrhenius theory. Even though each of the theories of chemical reaction has one or two drawbacks, they complement one another in explaining how chemical reactions take place. Transition state theory takes into account thermodynamic functions which the collision theory and the Arrhenius theory neglected. Because photocatalytic reactions are initiated and enhanced by light absorption, collision theory has relatively little direct explanation to offer. On the other hand, transition state theory attempts the theoretical explanation of the mechanism of chemical reaction and the calculation of absolute reaction rate. The theory explains the configurational change that occurs during chemical transformations which is accompanied by changes in potential energy in order to form products. The transition state theory assumes that all reactions proceed through the transition complex or activated complex and that the reactants are in quasi-equilibrium with the activated complex. The activated complex corresponds to the saddle in the plot of potential energy versus reaction progress (Fig. 3.1). It would be noted that the energy level of the activated complex in the photocatalytic reaction is lower than that of the uncatalysed reaction.

In photocatalytically favourable system, the reactant follows the traditional five steps involved in catalytic transformations with little modification based on the

U. I. Gaya, *Heterogeneous Photocatalysis Using Inorganic Semiconductor Solids*, DOI 10.1007/978-94-007-7775-0_3,

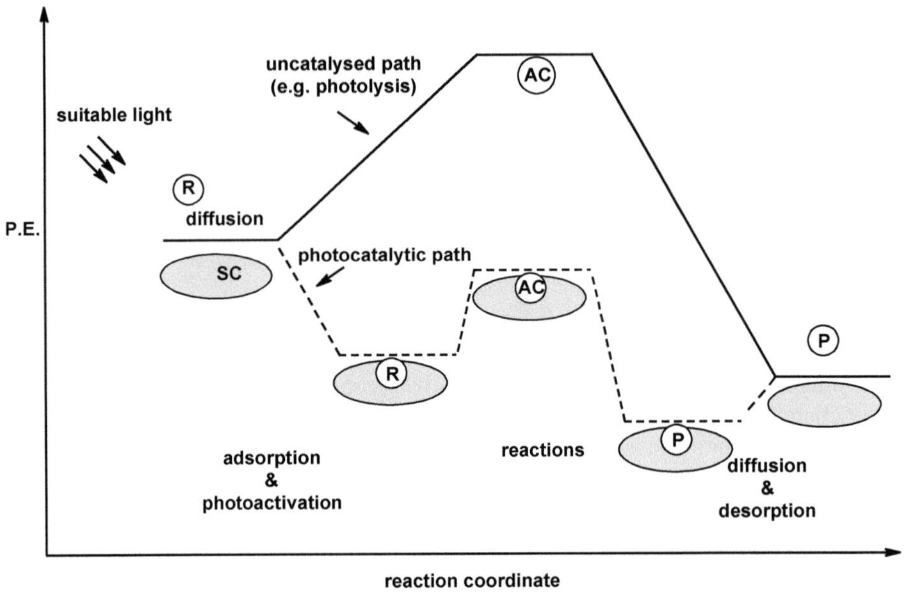

Fig. 3.1 Energetics depicting changes in potential energy (P. E.) in the conventional steps of photocatalytic reaction compared with uncatalysed process. *R* reactant, *AC* activated complex, *SC* semiconductor particle, *P* product(s)

incorporation of photoactivation. As seen in Fig. 3.1 the first step involves the diffusion of the reactant molecule towards the semiconductor photocatalyst where it gets adsorbed. The semiconductor photocatalyst undergoes photoexcitation by the absorption of band gap light to produce electron and hole. This step is crucial to any photocatalytic process. The charge carriers produced will get involved in transfer reactions. At this stage the activated complex is formed where bond formation and bond dissociation simultaneously occur as well as changes in functionalities. The rate of the photocatalytic reaction is proportional to the number of transition complexes that cross the activated complex barrier per unit time. The potential energy difference between the reactants and the transition complex corresponds to the activation energy E^* of the reaction. E^* is the minimum energy required to cause chemical reaction. In the final stage, the adsorbed products formed undergo desorption and diffuse into the fluid phase.

The truly catalytic nature of photocatalysis manifests when the conversion efficiency exceeds significantly the threshold conversion of the catalytic process. In order to conclude on the truly catalytic nature of a photocatalytic process the turnover number (TON; defined as the number of the reacting molecules or product molecules formed per surface active sites) must be known for both the photocatalytic reaction and for the catalytic reaction. If the ratio of the former to the latter is much greater than unity 1 the process is truly catalytic.

3.2 Langmuirian Chemical Adsorption Model

During heterogeneous photocatalytic reactions one or more of the reactants get adsorbed on a portion of the photocatalyst. In catalysis, this portion of adsorbent is known commonly as the active site. As photocatalysis is driven by light-induced charge carrier separation not by the activity of photocatalyst sites we will only adopt the use of photocatalyst site in this text to represent any portion of the semiconductor that adsorbs the reactant effectively in the presence of incident band gap light. There is great number of photocatalyst sites spread over the surface of a photocatalyst depending upon its nature.Basically, the fundamental theory of adsorption that applies to photocatalysis is the Langmuir theory which postulated that adsorption is limited to monolayer and excess molecules are therefore reflected back. The adsorption of chemical substances on the surface of photocatalyst is Langmuirian. That is to say, adsorption at one photocatalyst site does not affect the adsorption at adjacent site. The rate of adsorption of the molecules at equilibrium is equal to the rate of desorption. If A is adsorbed on a photocatalyst site S to form AS, the adsorbed phase, the process can be reversed to form the initial reactants A and S.

$$\mathrm{A} + \mathrm{S} \underset{k_{-1}}{\overset{k}{\rightleftharpoons}} \mathrm{AS} \tag{3.1}$$

AS and S can therefore be a measure of occupation and emptiness of surface sites respectively. This state of balance between the occupied photocatalyst site and the unoccupied site is called adsorption equilibrium. The equilibrium constant of adsorption may be derived from the rate of change in the concentration of occupied sites.

At equilibrium:

$$k_1[\mathrm{A}][\mathrm{S}] = k_{-1}[\mathrm{AS}] \tag{3.2}$$

$$\mathrm{K}_{\mathrm{ads(A)}} = \frac{k_1}{k_{-1}} = \frac{[\mathrm{AS}]}{[\mathrm{A}][\mathrm{S}]} \tag{3.3}$$

$\mathrm{K}_{\mathrm{ads(A)}}$ is the adsorption equilibrium constant.

We can deduce the fractional coverage at equilibrium θ_A as follows. The concentration of free photocatalyst potions [S] is related to the concentration of covered photocatalyst sites [AS] and the total concentration of catalyst sites $[S_T]$ the by the Eq. (3.4).

$$[\mathrm{S}] = [\mathrm{S_T}] - [\mathrm{AS}] \tag{3.4}$$

$$\Rightarrow \quad k_1\,[\mathrm{A}]\,([\mathrm{S_T}] - [\mathrm{AS}]) = k_{-1}\,[\mathrm{AS}] \tag{3.5}$$

So

$$k_1[\mathrm{A}][\mathrm{S_T}] - k_1[\mathrm{A}][\mathrm{AS}] = k_{-1}[\mathrm{AS}] \tag{3.6}$$

$$\Rightarrow \quad k_1\,[\mathrm{A}]\,[\mathrm{S_T}] = k_1\,[\mathrm{A}]\,[\mathrm{AS}] + k_{-1}\,[\mathrm{AS}] \tag{3.7}$$

Or

$$k_1[\mathrm{A}][\mathrm{S_T}] = (k_1[\mathrm{A}]+k_{-1})[\mathrm{AS}] \tag{3.8}$$

$$\therefore \quad \frac{k_1[\mathrm{A}]}{k_1[\mathrm{A}]+\mathrm{k}_{-1}} = \frac{[\mathrm{AS}]}{[\mathrm{S_T}]} \tag{3.9}$$

But

$$[\mathrm{AS}]/[\mathrm{S_T}] = \theta_\mathrm{A} \tag{3.10}$$

$$\therefore \quad \theta_\mathrm{A} = \frac{k_1[\mathrm{A}]}{k_1[\mathrm{A}]+k_{-1}} \tag{3.11}$$

By dividing the numerator and the denominator by k_{-1} we obtain the Langmuir equation:

$$\theta_\mathrm{A} = \frac{\mathrm{K_{ads(A)}}[\mathrm{A}]}{\mathrm{K_{ads(A)}}[\mathrm{A}]+1} \tag{3.12}$$

$\mathrm{K_{ads(A)}}$ is the adsorption equilibrium constant for a reactant A which could be obtained from the ratio of the rate constants $\frac{k_1}{k_{-1}}$.

Note: We can utilise a different assumption of the Langmuir theory to arrive at Eq. (3.12). Since Langmuir assumes that the surface of adsorbent is a uniform monolayer and that there will be no more adsorption once monolayer is covered up:

$$\text{rate of desorption of A} = \text{rate of adsorption of A} \tag{3.13}$$

So,

$$k_{-1}\theta_\mathrm{A} = k_1[\mathrm{A}](1-\theta_\mathrm{A}) \tag{3.14}$$

$$\theta_\mathrm{A} = \frac{k_1[\mathrm{A}]}{k_1[\mathrm{A}]+k_{-1}} \tag{3.15}$$

⇒ Because $\mathrm{K_{ads(A)}} = k_1/k_{-1}$

$$\theta_\mathrm{A} = \frac{\mathrm{K_{ads(A)}}[\mathrm{A}]}{\mathrm{K_{ads(A)}}[\mathrm{A}]+1} \tag{3.16}$$

$$\frac{1}{\theta} = \frac{1}{\mathrm{K_{ads(A)}}[\mathrm{A}]}+1 \tag{3.17}$$

The formula in Eq. 3.17 is known as the Langmuir isotherm because temperature is fixed (Mortimer 2008). The plot of $1/\theta$ versus $1/[\mathrm{A}]$ gives a straight line with slope equal to unity. The value of $1/\mathrm{K_{ads(A)}}$ corresponds to the concentration of A for which $\theta = 1/2$.

The reaction of adsorbed molecules with the surface is referred to as chemisorption. Practically, chemisorption does not go beyond monolayer. As Langmuir model predicts heat of adsorption that is independent of fractional coverage (Castellan 1983), the model can adequately describe chemisorptions processes including photocatalytic processes.

3.3 Rate Expressions for Heterogeneous Photocatalysis

In photocatalysis, two major processes dominate the background, mainly adsorption and irradiation. At the initial stage, the substrate molecule A diffuses to the surface of the semiconductor photocatalyst for adsorption as shown in Eq. (3.18).

$$A + SC \underset{k_{-1}}{\overset{k_1}{\rightleftharpoons}} A_{ads} \tag{3.18}$$

The rate of adsorption of substrate the molecule may be given as

$$r_{ads} = k_1[A] \tag{3.19}$$

Meanwhile, the semiconductor photocatalyst undergoes excitation to produce charge carriers as shown in Eq. (3.20).

$$SC \underset{k_{-2}}{\overset{k_2}{\rightleftharpoons}} SC^*(e^- + h^+) \tag{3.20}$$

In cases of poor charge separation, recombination takes place. The rate of recombination may be given by:

$$r_{rec} = k_{-2}[e^-][h^+] \tag{3.21}$$

Or

$$r_{rec} = k_{-2}[e^-]^2 \tag{3.22}$$

In absence of recombination, electron-hole transfer processes will take place to form products.

$$SC^*(e^- + h^+) + A_{ads} \xrightarrow{k_3} \text{Products} \tag{3.23}$$

For successful photocatalytic process, the rate of charge carrier generation (r_{gen}) must dominate sufficiently the rate of recombination (r_{rec}) such that some charge carriers escape and participate in electron/hole transfer. When first order rates are assumed (for convenience) for both mutual recombination and surface reaction of e^- and h^+ with the substrates, the rate of survival of electron and hole (r_{surv}) may be given, using steady-state approximation for charge carriers (Ikeda et al. 2001), as in Eq. 3.21. The rate should be constant and independent of whether the semiconductor photocatalyst is doped or undoped.

$$r_{surv} = \frac{I\phi k_{surv}[A]}{k_{surv}[A] + k_{-2}} \tag{3.24}$$

where k_{surv} is the rate constant for the survival of charge carriers, [A] is the surface concentration of substrate A, k_{-2} is the rate of recombination, I is the excitation photon flux and ϕ is the probability of light absorption to produce charge carriers. The overall rate of reaction is presumably directly proportional to r_{surv}. So, if recombination is negligible $k_{-2} \ll kC$ which means $r_{surv} = I\phi$. This implies maximum quantum yield. So far, it has been impracticable to have 100 % quantum yield in photocatalysis (Herrmann 2010a, b).

Generally, it is believed that the rate-controlling step in photocatalytic degradation is the photoabsorption of light and Langmuir model has been rigorously obeyed. The oft-misunderstood fact is that the mechanistic envision of Langmuir of uniformity of surface and independence of adsorption-desorption on adjacent photocatalyst sites may not hold true in real cases. Moreover, the kinetics of adsorption becomes more complicated than Langmuir since molecules can diffuse on the surface of photocatalysts (Engel and Reid 2006). In real heterogeneous photocatalytic systems, the application of Langmuir scheme to the rate of the reaction depends upon whether the overall reaction is controlled by diffusion or light (Ohtani 2010) as described below.

Photocatalytic-Controlled Process If the rate of surface reactions by photogenerated electrons and positive holes is slower than the adsorption of solutes, the surface concentration is always in adsorption equilibrium, and the overall rate becomes dependent on light-driven surface reactivity. For such a photocatalytic process the rate-limiting step is light-controlled. This is not discounting the fact that the surface reactions are still affected by physical parameters such as pH and ionic strength.

Given the presence of the substrate and oxidant, the initial rate of the photocatalytic process r will be given by the Langmuir-Hinshelwood bimolecular scheme:

$$\mathrm{r} = k_3 \theta_A \theta_B \tag{3.25}$$

where θ_A and θ_B are the fractional surface coverages of the substrate A and the active oxidising species B. The parallel adsorption of an oxidising species B can be represented by the following chemical equation.

$$\mathrm{B} + \mathrm{Photocatalyst} \rightleftharpoons \mathrm{B_{ads}} \tag{3.26}$$

The yield of adsorbed hydroxyl radical ($HO^{\bullet}{}_{ads}$) and H_2O_2 in irradiated aqueous TiO_2 suspension has been determined to be nearly constant across a wide pH range (Gao et al. 2002; Du and Rabani 2003). Thus, in pure solution of the substrate A, assuming constant light intensity and oxygen activity, $\theta_B = 1$ or constant. Hence, the initial rate of photocatalytic process can be given as:

$$\Rightarrow \quad \mathrm{r} = k_3 \theta_A \tag{3.27}$$

$$\therefore \quad \mathrm{r} = k_3 \frac{\mathrm{K_{ads(A)}[A]}}{\mathrm{K_{ads(A)}[A]} + 1} \tag{3.28}$$

It follows from the foregoing that at equilibrium Langmuir isotherm equation is consistent with the Langmuir-Hinshelwood equation and this has been established in myriad studies of photocatalytic destruction of organic substances (Horváth 2003; Dung et al. 2005; Valente et al. 2006; Priya and Madras 2006). Normally, at levels of $\mathrm{mmoldm^{-3}}$ the $[A] \ll 1$ and the Eq. (3.23) approximates the apparent first order rate equation (Konstantinou and Albanis 2004):

$$\mathrm{r} = k_3 \mathrm{K_{ads(A)}[A]} \tag{3.29}$$

Thus under the concentration condition mentioned above, the initial degradation rate could be fitted into the apparent first order rate law:

$$r = k_{app}[A] \tag{3.30}$$

Therefore, for molecules of A undergoing photocatalytic degradation

$$-\frac{d[A]}{dt} = k_3 K_{ads(A)}[A] \tag{3.31}$$

or

$$-\frac{d[A]}{[A]} = k_3 K_{ads(A)} dt \tag{3.32}$$

We can integrate both sides of Eq. (3.27) from initial limits of irradiation time and concentration ($t = 0$; $[A] = [A]_0$) to final limits of the variables ($t = t$; $[A] = [A]_t$)

$$-\int_{[A]_o}^{[A]_t} \frac{d[A]}{[A]} = k_3 K_{ads(A)} \int_{t=0}^{t=t} dt \tag{3.33}$$

$$\ln \frac{[A]_0}{[A]_t} = k_3 K_{ads(A)} t \tag{3.34}$$

$$\therefore \quad [A]_t = [A]_0 e^{-t.k}{}_{app} \tag{3.35}$$

where K_{app} is the apparent first order rate constant given by the slope of the graph of $\ln[A]_0/[A]_t$ versus t and $[A]_0$ is the initial concentration of the organic pollutant.

Based on Eq. (3.35) the disappearance of single substrate will be expected to agree with the first-order exponential decay model. Put simply, observation has shown that the apparent first-order rate equation can describe successfully the photodegradation of numerous organic compounds (De Heredia et al. 2001; Lachheb et al. 2002; Guettaï and Amar 2005; Chen et al. 2007; Caliman et al. 2007) even if it could not directly give adequate fitting (Demeestere et al. 2004). It is possible nonetheless, especially at saturation coverage to come across consistence with pseudo-zeroeth order kinetics (Dijkstra et al. 2001).

Considering the fact that photocatalytic reactions proceed through the concomitant formation of several intermediates with similarities in chemical behaviours, Eq. (3.24) can be extended by rate summation (Ollis 2005).

Thus,

$$r = k_3 K_{ads(A)}[A](1 + K_{ads(B)}[B] + K_{ads(C)}[C] + \ldots.. K_{ads(N)}[N]) \tag{3.36}$$

The terms representing the rates of the intermediates may vary negligibly compared to $K_{ads(A)}[A]$ or $1 \gg \sum_{N=B}^{N=N} K_{ads(N)}[N]$. Therefore, in such a case the following relationsip holds:

$$r = k_{app}[A] \tag{3.37}$$

Diffusion-Controlled Process When substrates are adsorbed on the surface of the semiconductor photocatalyst in solution, the process that restores the initial concentration at the sub-surface by transfer of more solutes from the bulk solution is known as diffusion. With increasing surface coverage, there will be slowing down of adsorption. If the rate of adsorption of the substrate from fluid phase is slower than the rate of electron-hole reactions with the substrate then diffusion prevails and the photocatalytic process becomes diffusion-controlled.

In diffusion-controlled processes, it is assumed there is no activation energy barrier to transfer solute to the surface (Chang and Franses 1995). Hence diffusion is the only mechanism to establish adsorption equilibrium in diffusion-controlled processes. Where this physical process becomes important it may result in disguise of the observed kinetics of photocatalytic reaction. Mass transfer (of solute from sub-surface to the bulk fluid) and diffusion influences may be observed under both steady and periodic illumination, in both liquid and gas phase photocatalytic systems. For example, diffusion enhanced gas-phase photocatalytic oxidation of trichloroethene has been observed in tubular-annular reactor upon controlled periodic illumination of TiO_2 films (Buechler et al. 1999).

The following equations apply to laminar flow process, in a tube with non-porous photocatalytic walls (Chen et al. 2000; Zhou and Ray 2003).

For first-order diffusion of A onto a semiconductor photocatalyst,

$$\text{rate of mass transfer r} = \text{rate of surface reaction} \tag{3.38}$$

$$k_m\left([\mathrm{A}]_\mathrm{b} - [\mathrm{A}]_\mathrm{s}\right) = k_s[\mathrm{A}]_\mathrm{s} \tag{3.39}$$

where k_m is the fluid to surface mass transfer, K_s is the rate constant for surface reactions, $[A]_s$ is the surface concentration and $[A]_b$ is the bulk concentration. k_m depends upon fluid velocity, fluid density, viscosity, tube diameter and molecular diffusivity. If we make $[A]_s$ the subject of the formula, then:

$$[\mathrm{A}]_\mathrm{s} = k_m[\mathrm{A}]_\mathrm{b}/(k_s + k_m) \tag{3.40}$$

So that

$$\mathrm{r} = k_\mathrm{s}[\mathrm{A}]_\mathrm{s} = \frac{[\mathrm{A}]_\mathrm{b}}{\left(\frac{1}{k_\mathrm{m}}+\frac{1}{k_\mathrm{s}}\right)} \tag{3.41}$$

Therefore

$$\mathrm{r} = k_\mathrm{app}[\mathrm{A}]_\mathrm{s} \tag{3.42}$$

Mass transfer limitations in photocatalytic reactors are measured mathematically by the ratio of surface concentration of reactant $[A]_s$ to the bulk concentration $[A]_b$, which means no mass transfer for a ratio of 1. So far we have discussed the effect of external mass transfer. Internal mass transfer within the porous parts of the semiconductor photocatalyst may also influence reaction rate. Internal mass transfer depends on the nature of the photocatalyst. For example, thin film photocatalyst can be considered a porous slabs and their internal mass transfer for first order photocatalytic process can be given by the famous Thiele modulus (Zhou and Ray 2003) shown in Eq. 3.39.

$$\Phi = H\sqrt{\frac{k}{D}} \tag{3.43}$$

Where Φ is the Thiele modulus, H is the thickness of the film, k is the first-order rate constant and D is the effective diffusivity of the pollutant within the photocatalyst. The lower the value of Φ the lower the effect of internal mass transfer. This happens as the first-order rate scheme collapses to zero order (due to surface saturation for example).

3.4 Pathway Intermediates, Mechanism and Validation

Reaction mechanism requires partly the description of what reactants and how they are converted to products at molecular scale. In photocatalytic processes, intermediates occur in varied forms from radicals, to inorganic ions to neutral organic species. There are several methods to detect and reveal these intermediates as they are formed. We attempt the discussion of these methods in the frame of two principal approaches: direct and indirect elucidation of intermediates.

3.4.1 Direct Detection of Transient Molecules and Radicals

3.4.1.1 Paramagnetic Resonance Methods

Reactive oxidising species involved in photocatalytic oxidation such as $^{\bullet}OH$, $O_2^{-\bullet}$, may have significant stability at room temperature (Yu et al. 2004). These reactive intermediates contain unpaired electrons and their generation can therefore be monitored by electron paramagnetic resonance using N-oxide spin trapping agents such as 5,5-dimethyl-1-pyrroline N-oxide (DMPO) or 4-pyridyl-1-oxide-N-tert-butylnitrone (POBN) as indicator (Schwarz et al. 1997; Brežová et al. 2009). We stress the use of the term electron paramagnetic resonance (EPR) as it is more encompassing than electron spin resonance (ESR) and it is not restricted only to spin systems. The EPR is a powerful technique based on the observation of the resonance magnetic field arising from the interaction of electromagnetic radiation with magnetic moments (from unpaired electrons). The electron possesses a magnetic moment μ with spin S = 1/2. Magnetic moment μ is allowed two orientations in presence of applied external magnetic field B, owing to Zeeman effect. The difference in energy between the two electron spin levels is encapsulated by

$$\Delta \mathrm{E} = g\mu_B B \tag{3.44}$$

where g is the electron g-factor, a dimensionless scalar, and μ_B is the Bohr's magneton $eh/4\pi m_e$ (a constant = 9.274 × 10^{-28} JG). The magnetic field B is measured in Tesla (T), where T = 10^4 gauss. As magnetic field is varied, absorption occurs at frequency of the oscillating magnetic field v (GHz or MHz). The energy of the absorption ΔE can be related to the magnetic flux density by Eq. (3.45).

$$\Delta \mathrm{E} = hv = g\mu_B B \text{ (assuming no hyperfine interactions)} \tag{3.45}$$

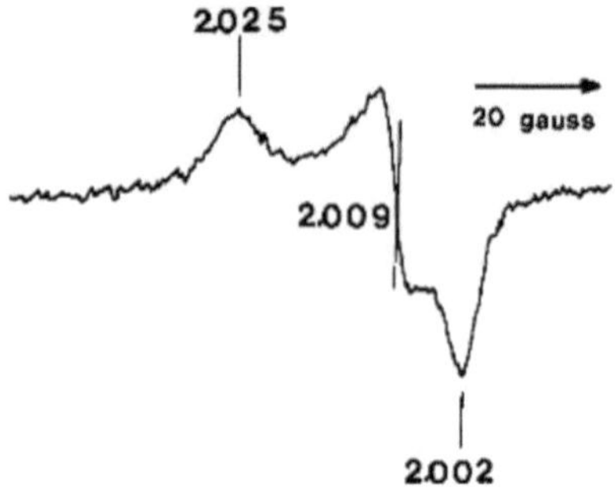

Fig. 3.2 EPR spectra of colloidal TiO_2, solution at pH 2.2 containing mol of 10^{-3} mol dm^{-3} H_2O_2 after irradiation at 77 or 4.2 K. (Reproduced from Howe and Grätzel 1985)

Paramagnetic species can exhibit anisotropy. In such a case they will be oriented in three directions in three spatial directions represented by 3 × 3 matrix g. Along x, y and z directions with respect to magnetic field the g-matrix factor can be denoted for example as g_x, g_y, g_z. In axial symmetry, g_x can be denoted by $g_{\|}$ whereas in the unaxial case g_y and g_z can be represented by $g_{\perp}$. In a spectrum having the characteristics of lower than axial symmetry with three distinct lines, g_1, g_2, and g_3 are used for the low, middle and high field line in that order.

Howe and Grätzel (1985) observed for the first time the trapping of electrons in colloidal TiO_2. Figure 3.2 shows the only EPR signal detected (at $g_1 = 2.025$, $g_2 = 2.009$, $g_3 = 2.002$) when acidic colloid solutions containing H_2O_2 were irradiated at -196°C. The signal was lost on warming to room temperature and was not observed when solutions irradiated at room temperature were subsequently cooled at -196°C. In the presence of peroxide no EPR signals were observed at $g < 2$ due to electron trapping by the peroxide, and no blue colour developed on irradiation. Rather, the colloid solutions are a faint yellow colour in the presence of peroxide. The electron trapping can be shown as follows:

$$H_2O_2 + e^- \rightarrow {}^{\bullet}OH + {}^{-}OH \tag{3.46}$$

Electron trapping at Ti^{4+}, usually characterised by Ti^{3+} signal at $1.900 < g < 2.000$, was not observed in the presence of H_2O_2, a scavenger. EPR signal is specific for a given molecule and its unpaired electron which makes paramagnetic resonance spectroscopy a powerful tool for the detection of radical species. A high number of radical species such as ${}^{\bullet}CH_3$,${}^{\bullet}CH_2COOH$, $CH_3CO^{\bullet}$, trapped electron in transition metals (because it localises in their d orbital), reactive oxygen species, bulk and surface non-metals, have been elucidated directly by EPR (Wang et al. 2011) Recently, Fu et al. (2006) carried out EPR measurements that gave the first direct evidence that the active species ${}^{\bullet}OH$ and $O_2^{-\bullet}$ are responsible for the photodecomposition of 4-chlorophenol over N-doped TiO_2 under visible-light irradiation.

3.4.1.2 Transient Absorption Spectroscopy

Time-resolved techniques measure the response of a substance after a rapid perturbation (following a change in reactant concentration, for instance). There is variety of transient methods often combined with spectroscopy. Usually, kinetic properties such as rate constant and amplitudes of chemical transformations of physical state taking place in a material are determined by measuring time dependent relaxation as a new equilibrium state is attained. These transient techniques have provided very sensitive means of understanding the fundamentals of photocatalysis from charge carrier dynamics to electron/hole transfer reactions. In these techniques, suitable fraction of organic molecules is transferred into excited state accompanied by significant laser pulses whose absorption spectrum can be measured. The study of rapid decay of this absorption spectrum is known as the transient absorption spectroscopy.

Several transient absorption methods have been used to ascertain the nature of charge carrier trapping and dynamics. Some notable examples of investigations carried out in this context are summarised below. Lawless et al. (1991) showed for the first time using pulse radiolysis, that trapped hole has the same identity with surface-bound •OH radical and can therefore be treated equivalently. Subsequently in a femtosecond spectroscopic using SCN-, Colombo and Bowmann (1996) confirmed that for species adsorbed to semiconductor photocatalyst, hole-transfer reaction can successfully compete with the picosecond electron-hole recombination process. Further investigations were made by Bahnemann et al. (1997). Using microsecond time-resolved spectroscopic data (which includes trapped holes and their yield) to reveal details about electron transfer and other aspects of photocatalytic mechanism. A more recent transient absorption study was carried out by Yang and Tamai (2001) at femtosecond scale, to characterise trapped holes. The transient absorption spectra (Fig. 3.3) of colloidal systems of nanosized, anatase TiO_2 was compared with and without SCN^-. Broad absorption band around 520 nm was observed immediately after band-gap excitation for the system without SCN^- has been assigned to shallowly trapped holes. In the presence of SCN^- however, the absorption from the trapped holes at 520 nm cannot be observed because of the ultrafast interfacial hole transfer between nanoparticles and SCN^-. The SCN^- hole trapping can be represented by the following chemical equations:

$$h_{tr}{}^{+} + SCN^- \rightarrow SCN^{\bullet} \tag{3.47}$$

$$SCN^- + SCN^{\bullet} \rightarrow (SCN)_2{}^{\bullet -} \tag{3.48}$$

These findings reinforce the 450 nm band obtained by Bahnemann et al. (1997) in a different study at microsecond scale. The times of hole and electron trapping were estimated to be < 50 and 260 fs, respectively, by the analysis of rise and decay dynamics of transient absorption spectra.

3.4.1.3 Coupled or Single Chromatography

Most organic and even inorganic products of photocatalytic degradation are separable on chromatographs. Chromatographic methods frequently used are mainly liquid-based versions such as high performance liquid chromatography (HPLC) and gas chromatography. The facilities are installed with suitable column and detector.

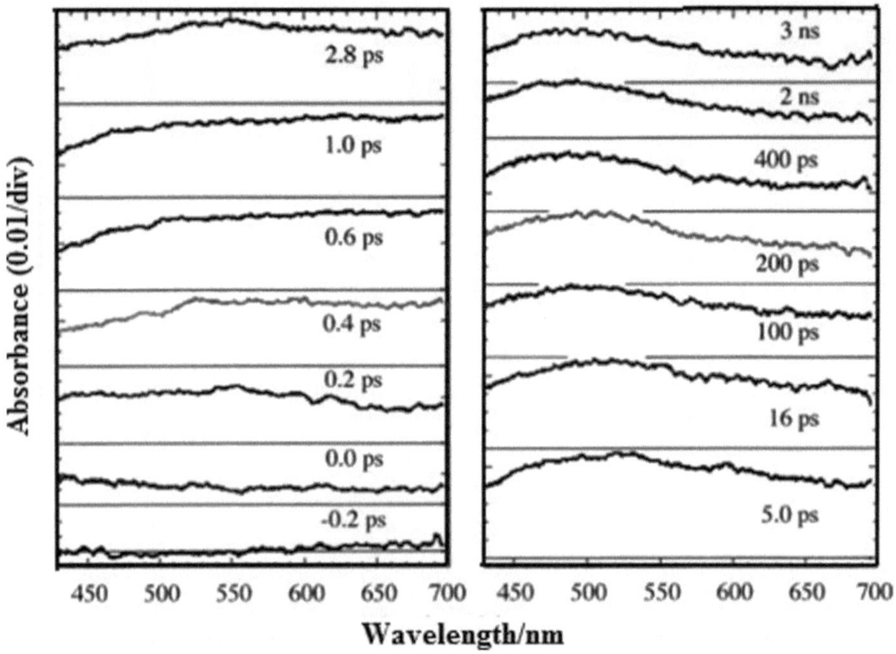

Fig. 3.3 Transient absorption spectra of colloidal aqueous solution of TiO_2 (6 gdm^{-3}) with adsorbed SCN^- (0.3 mol dm^{-3}) at various time delays (shown on the figure) after excitation at 360 nm probed by a femtosecond white-light continuum. (Reproduced from Yang and Tamai 2001)

Liquid chromatographic methods are usually non-destructive and can separate component solutes of a mixture present in solution of ultratrace to major concentrations (Robinson et al. 2005). These methods can analyse the total aromatic intermediates in a photocatalytic system (Dieckmann and Gray 1996). Even though UV-detectors can be applied as detectors for liquid chromatographs, mass spectrometers perform unrivalled detection and offer higher predictive powers. Mass spectrometers such as time-of-flight types ensure higher resolution and mass accuracy (Niessen 2001). Such mass spectrometers come with analysers that give high specificity and allow the assignation of the empirical formula for unknown intermediate species of photocatalytic degradation including pharmaceutical intermediates (Sirtori et al. 2009). Mass spectrometer libraries (such as that of the Wiley or National Institute of Standards and Technology, NIST) have been helpful in the assignment of transition structures. Besides library searching, tentatively identified compound may be confirmed by running external standards. Matching structures with standards makes this method direct, thereby eliminating the need for complementary techniques.

In dealing with substituted intermediates, direct detection of de-substituted moiety would imply scission of the substituent group. For example, based on the direct detection of –OH as substituent group attached to intermediates of photocatalytic degradation, it is believed that the reaction scheme that accounts for the formation these intermediates is •OH attack or hydroxylation. Consequently, the

Table 3.1 Intermediates of 4-chlorophenol degradation over semiconductor oxide photocatalysts and pathways suggested by different investigators

Compound	Intermediate	Pathway	Reference
4-chlorophenol	Hydroquinone	I & II	(Theurich et al. 1996)
	Pyrocatechol	I & III	(Gaya et al. 2009)
	5-chloro 2,4′-dihydroxybiphenyl	I & IV	(Oudjehani and Boule 1992)
	4-chlorobenzene-1,2-diol	III	(Mills et al. 1993)
	4-chlorobenzene-1,3-diol	III	
	1,4-benzoquinone	V	
	Phenol	VI	(Theurich et al. 1996)

retrogressive steps of the photocatalytic degradation and finally a tentative mechanism can be surmised. The previous example is encountered when dealing with photocatalytic degradation of phenolic rings. A typical photo-oxidation mechanism for 4-chlorophenol based on the formation of pathway intermediates displayed in Table 3.1 is shown in Fig. 3.4. This compound undergoes several primary and secondary photoprocesses represented in both the table and figure by roman letters I-VI which include dechlorination (e.g. reaction I), para- or ortho- hydroxylation (e.g. reactions II and III), dimerisation (e.g. reaction IV), tautomerism (e.g. reaction V) and dehydroxylation (e.g. reaction VI).

Eventually, most aromatics including the phenolics form hydroquinones which transform to mainly benzoquinones (which are chemically non-aromatics), then to aliphatic compounds before the formation of CO_2 and H_2O.

3.4.1.4 Total Organic Carbon

One of the most important direct detection techniques is the total organic carbon (TOC) analysis as it measures photocatalytic mineralisation of organic pollutants. The significance of TOC measurement is that the final products of photocatalytic degradation may not always be purely benign and leftovers can be toxic as well or even more toxic (Oudjehani and Boule 1992). Total organic carbon can be measured using TOC analysers. Unlike high performance liquid chromatography, TOC analysis has the disadvantage of being non-specific. For organic compounds containing heterogroups that cannot be determined by TOC, the yield of the corresponding anions or radicals of the heteroatoms can be detected by suitable analytical methods (for example by HPLC equipped with conductivity detector) but this does not necessarily imply mineralisation owing to the possible presence of organics structures containing no heteroatoms in solution (Dai et al. 2008).

3.4.1.5 Quantitative Evidences

Certain quantitative parameters can allow drawing useful conclusions on reaction mechanism and this can be accepted as direct evidence depending on analytical

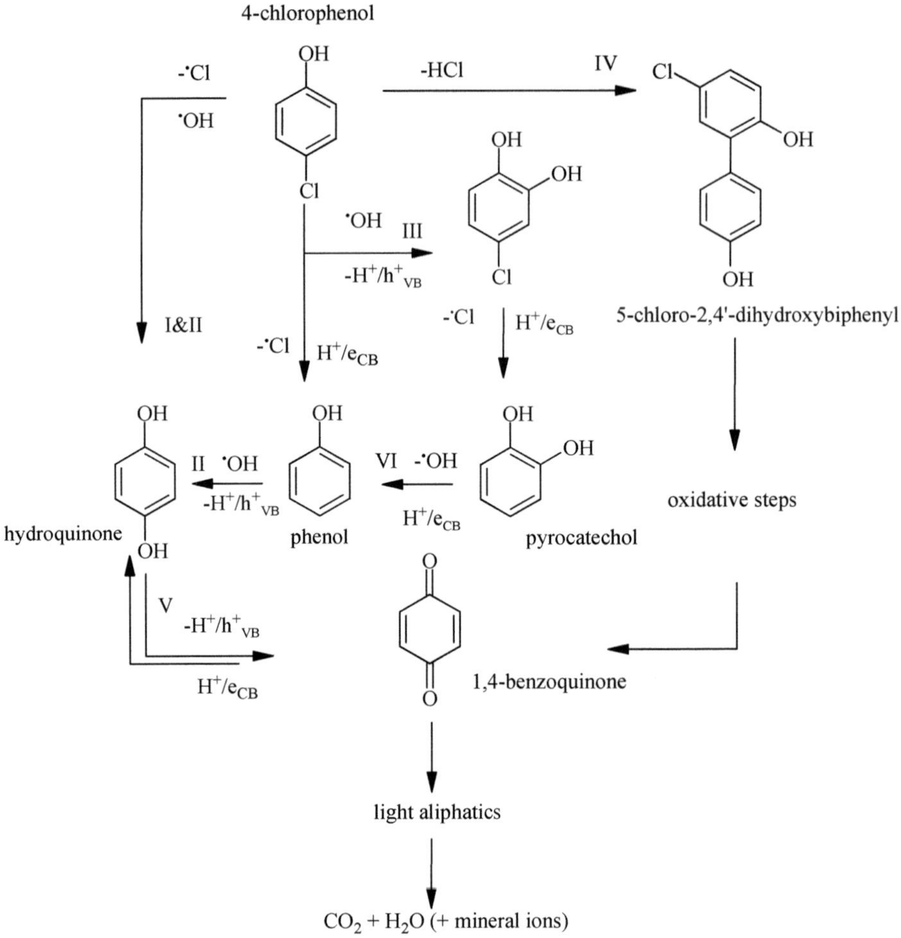

Fig. 3.4 Speculated mechanism for the photocatalytic degradation of 4-chlorophenol based on direct detection of intermediates. *I* p-declorination, *II* p-hydroxylation, *III* o-hydroxylation, *IV* dimerisation, *V* tautomerism, *VI* dehydroxylation

conditions. For example, different reactions can have the same half-life, as a result half-life cannot be used as a characteristic of a reaction. However, when the theoretical (or mechanistic) rate constant (or half-life) of an expected reaction is equal to the experimental rate constant (or half-life), that evidence may be considered acceptable. Similarly, Ishibashi et al. (2000) estimated the quantum yield of •OH production in aqueous solution during TiO_2 photocatalysis to be 7×10^{-5} by means of a method using terephthalic acid as a fluorescence probe and compared this quantum yield to that of hole generation estimated by iodide ion oxidation (5.7×10^{-2}) to arrive at the fact that the main oxidant in photocatalytic oxidation processes is hydroxyl radical.

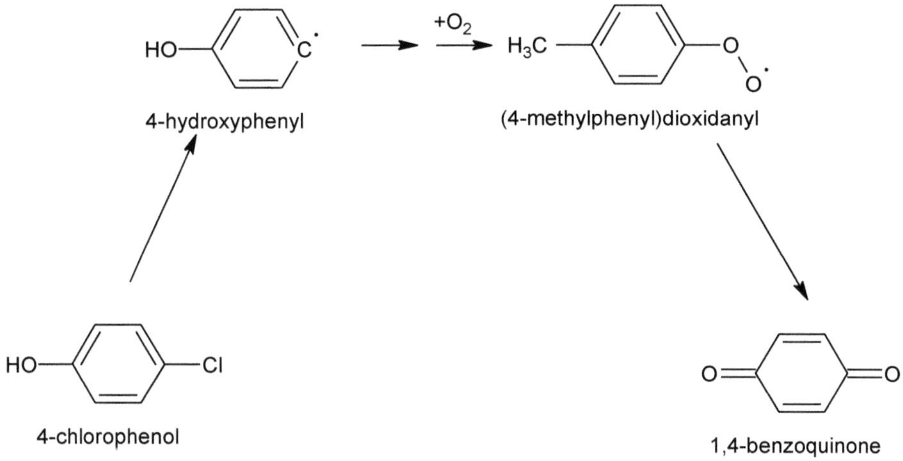

Fig. 3.5 Hydroxyphenyl pathway for the conversion of 4-chlorophenol to 1,4-benzoquinone

3.4.2 Indirect Elucidation of Intermediates

3.4.2.1 Based on Expected Chemistry

In establishing reaction mechanism, facts that are posited on well-known chemical basis can be accepted in absence of direct experimental account of observations. For instance, Theurich et al. (1996) proposed the mechanism for the formation of 1,4-benzoquinone from 4-chlorophenol to be via 4-hydroxyphenyl radical (Fig. 3.5). In real fact, this aryl radical was only determined by EPR technique, in the photolysis of aqueous 4-chlorophenol (Lipczynska-Kochany et al. 1991). This proposition is validated by the fact that the chemistry is expected to be the same since the starting material, the oxidant, the end product and reaction conditions are more or less the same for both advanced oxidation processes.

3.5 Conclusion

The trend in analytical chemistry is to improve performance as regards sensitivity, accuracy, precision and speed of operation. These developments often move in parallel with progress in existing fields of research, which photocatalysis is not an exception. The extent to which the elucidation of mechanisms of photocatalysis can reach will continue to depend on the robustness and sophistication of instrumental methods for application. Currently, coupled chromatography-mass spectrometry appears unrivalled in the determination of long-lived organic intermediates of photocatalytic reactions. Short-lived intermediates and oxidising species are determined by complementary methods. Future developments will surely rely on the possible improvement in the ability of existing analytical instruments.

References

Bahnemann DW, Hilgendorff M, Memming R (1997) Charge carrier dynamics at TiO_2 particles: reactivity of free and trapped holes. J Phys Chem B 101:4265–4275

Brežová V, Vrecková Z, Billik P et al (2009) Photoactivity of mechanochemically prepared nanoparticulate titanium dioxide investigated by EPR spectroscopy. J Photochem Photobiol A 206:177–187

Buechler KJ, Noble RD, Koval CA (1999) Investigation of the effects of controlled periodic illumination on the oxidation of gaseous trichloroethylene using a thin film of TiO_2. Ind Eng Chem Res 38:892–896

Caliman AF, Cojocaru C, Antoniadis A (2007) Optimized photocatalytic degradation of Alcian Blue 8 GX in the presence of TiO_2 suspensions. J Hazard Mater 144:265–273

Castellan GW (1983) Physical chemistry. Addison-Wesley Publishing Company, United States

Chang C-H, Franses EI (1995) Adsorption dynamics of surfactants at the air/water interface: a critical review of mathematical models, data, and mechanisms. Colloids Surf A 100:1–45

Chen D, Li F, Ray AK (2000) Effect of mass transfer and catalyst layer thickness on photocatalytic reaction. ALCHE 46:1034–1045

Chen J-Q, Wang D, Zhu M-X et al (2007) Photocatalytic degradation of dimethoate using nanosized TiO_2 powder. Desalination 207:87–94

Colombo DP Jr, Bowman RM (1996) Does interfacial charge transfer compete with charge carrier recombination? A femtosecond diffuse reflectance investigation of TiO_2 nanoparticles. J Phys Chem 100:18445–18449

Dai K, Peng T, Chen H et al (2008) Photocatalytic degradation and mineralization of commercial methamidophos in aqueous titania suspension. Environ Sci Technol 42:1505–1510

Davydov L, Smirniotis PG (2000) Quantification of the primary processes in aqueous heterogeneous photocatalysis using single-stage oxidation reactions. J Catal 191:105–115

De Heredia JB, Torregrosa J, Dominguez JR (2001) Oxidation of p-hydroxybenzoic acid by UV radiation and by TiO_2/UV radiation: comparison and modelling of reaction kinetic. J Hazard Mater 83:255–264

Demeestere K, De Visscher A, Dewulf J et al (2004) A new kinetic model for titanium dioxide mediated heterogeneous photocatalytic degradation of trichloroethylene in gas-phase. Appl Catal B 54:261–274

Dieckmann MS, Gray KA (1996) A comparison of the degradation of nitrophenol via direct and sensitised photocatalysis in TiO_2 slurries. Water Res 30:1169–1183

Dijkstra MFJ, Michorius A, Buwalda H (2001) Comparison of the efficiency of immobilized and suspended systems in photocatalytic degradation. Catal Today 66:487–494

Du Y, Rabani J (2003) The measure of TiO_2 photocatalytic efficiency and the comparison of different photocatalytic titania. J Phys Chem B 107:11970–11978

Dung NT, Khoa NV, Herrmann J-M (2005) Photocatalytic degradation of reactive dye RED-3BA in aqueous TiO_2 suspension under UV-visible light. Int J Photoenergy 7:11–15

Engel T, Reid P (2006) Physical chemistry. Pearson Education, San Fransisco

Fu H, Zhang L, Zhang S et al (2006) Electron spin resonance spin-trapping detection of radical intermediates in N-Doped TiO_2-assisted photodegradation of 4-chlorophenol. J Phys Chem B 110:3061–3065

Gao R, Stark J, Bahnemann DW et al (2002) Quantum yields of hydroxyl radicals in illuminated TiO_2 nanocrystallite layers. J Photochem Photobiol A 148:387–391

Gaya UI, Abdullah AH, Zainal Z et al (2009) Photocatalytic treatment of 4-chlorophenol in aqueous ZnO suspensions: intermediates, influence of dosage and inorganic anions. J Hazard Mater 168:57–63

Guettaï N, Amar HA (2005) Photocatalytic oxidation of methyl orange in presence of titanium dioxide in aqueous suspension. Part II: kinetics study. Desalination 185:439–448

Herrmann J-M (2010a) Fundamentals and misconceptions in photocatalysis. J Photochem Photobiol A 216:85–93

Herrmann J-M (2010b) Photocatalysis fundamentals revisited to avoid several misconceptions. Appl Catal B 99:461–468

Horváth IT (eds) (2003) Encyclopedia of catalysis. vol 5. J Wiley, New Jersey, pp 577–611
Howe RF, Grätzel M (1985) EPR observation of trapped electrons in colloidal TiO_2. J Phys Chem 89:4495–4499
Ikeda S, Sugiyama N, Pal B et al (2001) Photocatalytic activity of transition-metal-loaded titanium (IV) oxide powders suspended in aqueous solutions: Correlation with electron hole recombination kinetics. Phys Chem Chem Phys 3:267–273
Ishibashi K, Fujishima A, Watanabe T et al (2000) Quantum yields of active oxidative species formed on TiO_2 photocatalyst. J Photochem Photobiol A 134:139–142
Jardim WF, Moraes SG, Takiyama MMK (1997) Photocatalytic degradation of aromatic chlorinated compounds using TiO_2: toxicity of intermediates. Wat Res 31:1728–1732
Konstantinou IK, Albanis TA (2004) TiO_2-assisted photocatalytic degradation of azo dyes in aqueous solution: kinetic and mechanistic investigations: a review. Appl Catal B 49:1–14
Lachheb H, Puzenat E, Houas A et al (2002) Photocatalytic degradation of various types of dyes (Alizarin S, Crocein Orange G, Methyl Red, Congo Red, Methylene Blue) in water by UV-irradiated titania. Appl Catal B 39:75–90
Lawless D, Serpone N, Meisel D (1991) Role of OH radicals and trapped holes in photocatalysis. A pulse radiolysis study. J Phys Chem 95:5166–5170
Lipczynska-Kochany E, Kochany J, Bolton JR (1991) Electron paramagnetic resonance spin trapping detection of short-lived radical intermediates in the direct photolysis of 4-chlorophenol in aerated aqueous solution. J Photochem Photobiol A 62:229–240
Mills A, Morris S, Davies R (1993) Photomineralisation of 4-chlorophenol sensitied by titanium dioxide: a study of intermediates. J Photochem Photobiol A 70:183–191
Mortimer RG (ed) (2008) Physical chemistry. Elsevier Academic Press, Canada
Niessen WMA (2001) Current practice of gas chromatography-mass spectrometry. Marcel Decker, New York
Ohtani B (2010) Photocatalysis A to Z—what we know and what we do not know in a scientific sense. J Photochem Photobiol C 11:157–178
Ollis DF (2005) Kinetic disguises in heterogeneous photocatalysis. Top Catal 35(3–4):217–223
Oudjehani K, Boule P (1992) Photoreactivity of 4-chlorophenol in aqueous solution. J Photochem Photobiol A 68:363–373
Priya MH, Madras G (2006) Kinetics of Photocatalytic degradation of chlorophenol, nitrophenol, and their mixtures. Ind Eng Chem Res 45:482–486
Robinson JW, Frame EMS, Frame IIGM (2005) Undergraduate instrumental analysis. Marcel Dekker, New York
Schwarz PF, Turro NJ, Bossmann SH (1997) A new method to determine the generation of hydroxyl radicals in illuminated TiO_2 suspensions. J Phys Chem B 101:7127–7134
Sirtori C, Zapata A, Malato S et al (2009) Solar photocatalytic treatment of quinolones: intermediates and toxicity evaluation. Photochem Photobiol Sci 8:644–651
Theurich J, Lindner M, Bahnemann DW (1996) Photocatalytic degradation of 4-chlorophenol in aerated aqueous titanium dioxide suspensions: a kinetic and mechanistic study. Langmuir 12:6368–6376
Valente JPS, Padilha PM, Florentino AO (2006) Studies on the adsorption and kinetics of photodegradation of a model compound for heterogeneous photocatalysis onto TiO_2. Chemosphere 64:1128–1133
Wang Z, Ma W, Chen C et al (2011) Probing paramagnetic species in titania-based heterogeneous photocatalysis by electron spin resonance (ESR) spectroscopy—a mini review. Chem Eng J 170:353–362
Yang X, Tamai N (2001) How fast is interfacial hole transfer? In situ monitoring of carrier dynamics in anatase TiO_2 nanoparticles by femtosecond laser spectroscopy. Phys Chem Chem Phys 3:3393–3398
Yu J, Chen J, Li C (2004) ESR Signal of superoxide radical anion adsorbed on TiO_2 generated at room temperature. J Phys Chem B 108:2781–2783
Zhou S, Ray AK (2003) Kinetic studies for photocatalytic degradation of eosin B on a thin film of titanium dioxide. Ind Eng Chem Res 42:6020–6033

Chapter 4
Origin of the Activity of Semiconductor Photocatalysts

Abstract The properties that affect the photoactivity of semiconductor photocatalyst are pervasive and the interplay of variables in the semiconductor-assisted photoprocesses is complex. Understanding the heterogeneous, advanced oxidation photoprocess is crucial to its harvest, and must therefore be emphasized. This chapter is devoted to the highlight of critical properties necessary for the realisation of efficient photocatalytic process, cutting across surface science, solid state synthesis and process operation.

The properties that affect the photoactivity of semiconductor photocatalyst are pervasive and the interplay of variables in the semiconductor-assisted photoprocesses is complex. Understanding the heterogeneous, advanced oxidation photoprocess is crucial to its harvest, and must therefore be emphasized. This chapter is devoted to the highlight of critical properties necessary for the realisation of efficient photocatalytic process, cutting across surface science, solid state synthesis and process operation.

4.1 Background Sketch

The major component of a basic semiconductor photocatalytic process is the photocatalyst. Among semiconductors materials, pure or modified forms of TiO_2 and ZnO have shown superior applicability and versatility. In order to enhance the efficiency of these photocatalysts, many strategies have been adopted to form the right combination of surface morphology, local structure and crystal phase. The synthesis methods, being central in this context, have been revolutionised in the search for suitable catalyst choices for the purpose. Armed with the knowledge of basic solid state chemistry and physics, these synthesis techniques have provided several synthesis routes which have continually been exploited to provide desired semiconductor photocatalysts for different photocatalytic applications. Synthesis methods can be categorised into mainly, but not restricted to co-precipitation, combustion, vapour deposition, sol-gel and solvothermal techniques. Globally speaking, there are various physical techniques for the production of thin film and powder photocatalyst materials as we shall see in the following discussions.

U. I. Gaya, *Heterogeneous Photocatalysis Using Inorganic Semiconductor Solids*, DOI 10.1007/978-94-007-7775-0_4,

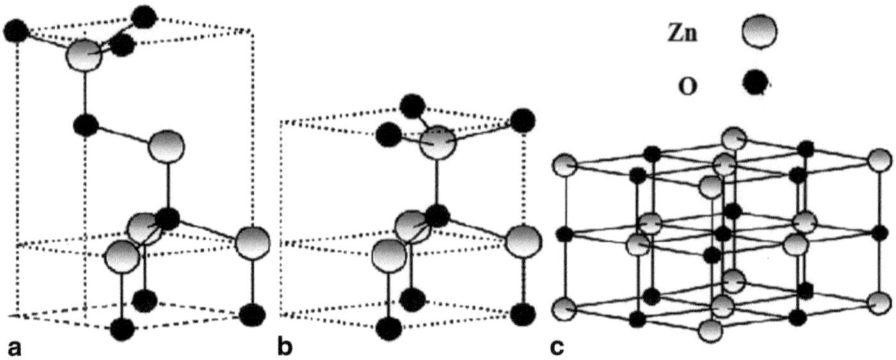

Fig. 4.1 Crystal structures of common ZnO phases: zinc blende (B3) (**a**), rocksalt (**b**), wurtzite (**c**). (Reproduced from Morkoç and Özgür 2009)

4.2 Photocatalytic Efficiency

4.2.1 Basic Determinants of Photocatalytic Efficiency

Semiconductor photoctalysis can be influenced by some many operational and non-operational factors. There are primary factors such as catalyst structure and composition, and the resulting processes that take place upon interaction with light. These properties may be regarded as basic determinants of photocatalytic performance. Prior to operation, basic such properties can be modified towards better photocatalytic activity. Some of these basic determinants of photocatalytic efficiency are highlighted below.

4.2.2 Structure and Phase Composition

The anatomy of semiconductor particles is an important determinant of photocatalytic activity. Semiconductor photocatalysts can be synthesized in different polymorphic forms. Zinc oxide occurs in crystal form as cubic zinc blende (B3), hexagonal wurtzite (B4) and cubic rocksalt (or Rochelle salt) (B1). A stick and ball representation of these crystal structures is shown in Fig. 4.1 (Morkoç and Özgür 2009). The wurzitic ZnO phase has been the most commonly exploited in photocatalysis. This phase crystallises at ambient conditions and has promising efficiency in photocatalytic applications.

The common polymorphs of TiO_2 are anatase, brookite, rutile, and TiO_2 (B). These have tetragonal, orthorhombic, tetragonal and monoclinic cell structure, respectively (Fierro 2006). The common crystal forms of TiO_2 are shown schematically in Fig. 4.2. Among the three most common phases of TiO_2, rutile has been identified as a stable form while anatase and brookite are metastable (Gouma and Mills

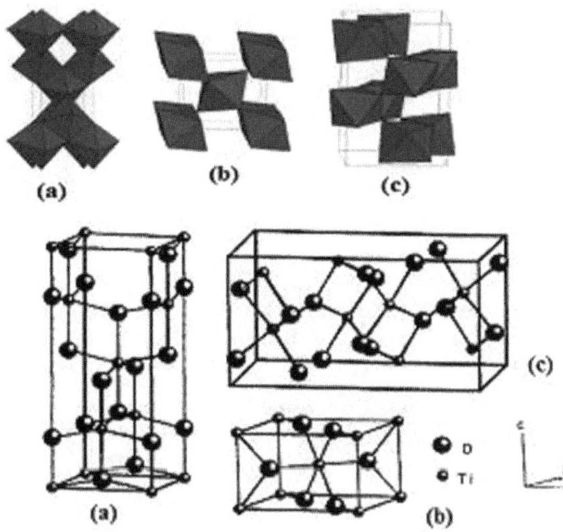

Fig. 4.2 Crystal structures of common TiO_2 phases: anatase (**a**), rutile (**b**), brookite, (**c**). (Reproduced from Mo and Ching 1995; Carp et al. 2004)

2001). On compression, titania forms high pressure polymorphs that are isostructural with orthorhombic α-PbO (columbite) and monoclinic ZrO_2 (baddeleyite) (Arlt et al. 2000; Dubrovinskaia et al. 2001). Even more polymorphs can be obtained by the further compression at ambient temperature or at elevated temperatures.

There are several commercial mixed-phase TiO_2 powders in the market with different trademark names such as Hombicat UV100 and PC500 (which are both 100 % anatase and have higher surface area) but Degussa P25 has outperformed most of these preparations and has therefore gained acceptance as a standard photocatalyst (Saquib and Muneer 2003). Degussa P25 is a nonporous, 70:30 % anatase-rutile mixture with a BET surface area of ~55 m^2g^{-1} and crystallite sizes of 30 nm in 0.1 pm diameter aggregates (Hoffmann et al. 1995). Bacsa and Kiwi (1998) have shown that the activity of these catalysts is not a function of surface area alone. The 100 % anatase sample which has a high surface area shows lower activity than the mixed phases of anatase and rutile, the highest activity standing at 30 % rutile. These findings inform the decision to regard the commercially available Degussa P25 as a standard, not discounting the fact that many titania preparations with higher photocatalytic activity have been reported. Nonetheless, some workers have synthesized single anatase or rutile TiO_2 phase with outstanding activity for a wide variety of photo-oxidation reactions, compared to the standard P25 (Sclafani and Herrmann 1996; Fujishima and Rao 1997). Similarly, with TiO_2 samples having the same surface area, the brookite nanoplates demonstrated the highest efficiency in the bleaching of methyl orange solution when compared with rutile nanorod and nanocrystalline anatase (Li et al. 2007).

Basically, anatase titania offers enhanced photocatalytic performance compared to rutile and its presence in TiO_2 powders has been shown to be greatly beneficial when used in photocatalysis. Several explanations have been given by different workers to account for this observation, which include the difference in the location of oxygen vacancies in the anatase and rutile energy gaps. In certain instances however, anatase and rutile have singly shown inefficiency in photocatalysis compared to the bi-phasic photocatalysts (Ohno et al. 2001a), what can be attributed to the larger specific surface area of anatase and variations in the rate constants for recombination (Riegel and Bolton 1995; Pekakis et al. 2006). It has been observed that the presence of a small amount of an anatase component is crucial for an efficient photocatalytic reaction on TiO_2 particles using oxygen as the electron acceptor (Ohno et al. 2001b; Emeline et al. 2002). Hurum et al. (2003) has revealed through ESR study with Degussa P25 that rutile is more integral to the activity of the catalyst than a simple electron sink for anatase. These workers explained that the smaller band gap of rutile extends the useful range of photoactivity into the visible region and stabilizes charge separation by electron transfer from rutile to anatase (or slows recombination) which is facilitated by the small size of the rutile crystallites.

Accordingly, mixing anatase, brookite and rutile nano-TiO_2 powders was found to yield remarkable photocatalytic performance for 4-nitrophenol photodegradation (Paola et al. 2008). Similarly, the photocatalytic efficacy of anatase-brookite composite was found to be greater than that of a single-phase anatase sample showing a much larger surface area (Ardizzone et al. 2007). However, other workers observed more photocatalytic hydrogen evolution activity of anatase/brookite mixtures and of pure brookite compared to pure anatase nanoparticles despite the lower surface area of the former but contrary behaviour was observed in the case of photocatalytic oxidation (Kandiel et al. 2010).

It is admittedly difficult to estimate the ratios of active and inactive phases in photocatalyst samples which led to the emphasis on crystallinity as a major indicator of photoactivity. Basically, high degree of crystallinity indicates low surface defects, low amorphous content and reduced the recombination rate. For this simple reason, the assessing the activity of amorphous component has been overlooked in many of the earlier studies. Kominami et al. (2002) have clarified that though the rate of photocatalytic reaction and the amount of substrates on the titania phases in their mixture cannot be discriminated, their different photoabsorption characteristics and the inefficiency of the amorphous phase is well known. Apart from inactivity, amorphous phases are either absent (Kao et al. 2007) or present in negligible quantities. An instance of the latter is that the content of amorphous particles in Degussa P25 TiO_2 has been observed by transmission electron microscopy (TEM) to be very low, only 1 % (Ohno et al. 2001a).

Normally, semiconductor photocatalysts occur largely as crystalline inorganic solids. The size and perfectness of the crystallite particles can both be derived from sharpness and shape of X-ray diffraction (XRD) peaks. The sharper the peaks the higher the crystallinity, and the lower the lattice defects and strain. Below μm level, crystallite size can be calculated from XRD data using Debye-Scherrer formula (Eq. 4.1).

$$G_{\mathrm{XRD}} = \frac{K\lambda}{\beta \cos\theta} \tag{4.1}$$

Where G_{XRD} is the crystallite particle size, λ is the X-radiation wavelength, K is the shape factor, θ is the Bragg's or diffraction angle, β is corrected line broadening at half the maximum intensity known as the full width at half maximum (FWHM). The most widely used correction for β is given by $\beta_s - \beta_o$, where β_s and β_o are the peak half width of sample and standard respectively. K is related to the crystallite size and has a typical value of 0.891 (Klug and Alexander 1974).

The crystal structure of semiconductor photocatalyst can affect its photoreactivity. Anatase titanium dioxide is characterised by the theoretically more reactive (001) surface and the thermodynamically morestable (101) plane. However, most of the natural and frequently synthesised anatase titanium dioxide powders for photocatalytic purposes is dominant in the (101) family of planes. Moreover, most of these preparations are polycrystalline despite the fact single crystals are known to provide special optoelectronic properties. From basic solid state chemistry, single crystal can be defined as a crystalline solid in which the crystal lattice is entirely continuous and unbroken to the edges of the solid sample. Recently, the synthesis of anatase TiO_2 single crystal has been demonstrated by Yang et al. (2009) along with the first experimental evidence on the reactivity of the predominantly (001) surface. Details of the synthesis will be given under solvothermal synthesis. The anatase TiO_2 showed 5 times the photoreactivity of Degussa P25 TiO_2. The transmission electron microscopy (TEM) images and selected area electron diffraction (SAED) pattern of this photoactive anatase material is given by Fig. 4.3. Parts of the Figure (Fig. 4.3a, c and d) confirm the single-crystalline nature of the anatase and indexed the SAED pattern as (001) plane. The high-resolution TEM image recorded from the white circled area in Fig. 4.3c depicts continuous (200) atomic planes with a lattice spacing of 1.9 Å (Fig. 4.3e) which corresponds to the {200} planes of anatase TiO_2 single crystals. Figure 4.3b is a schematic illustration of the crystal orientation of the anatase TiO_2 single crystals.

4.2.2.1 Surface Characteristics

Surface area determines the part of the photocatalyst which would be contacted by the substrate as well as the oxidising medium. The influence of surface area is partly dependent upon preparation method. Gaya et al. (2010) found that the performance of commercial ZnO from Merck, Alfa Aesar, PC Lab, which were presumably prepared by different methods, is surface area dependent. The Merck ZnO which has demonstrated high photoactivity is a non-porous material as shown by the adsorption isotherm in Fig. 4.4. On the other hand, Li and Heneda (2003) have shown that the photocatalytic activity of ZnO prepared by the same method depends on crystallinity rather than surface area. Over several decades, surface area has been measured based on the famous Brunauer-Emmett-Teller (BET) principle introduced by Brunauer et al. (1938). If we assume negligible porosity and surface roughness, BET surface area $A_{s(BET)}$ (m^2g^{-1}) can be related to crystallite particle size G_{BET} (nm) by the Eq. (4.2):

$$A_{s(BET)} = \frac{6x\ 10^3}{\rho G_{BET}} \tag{4.2}$$

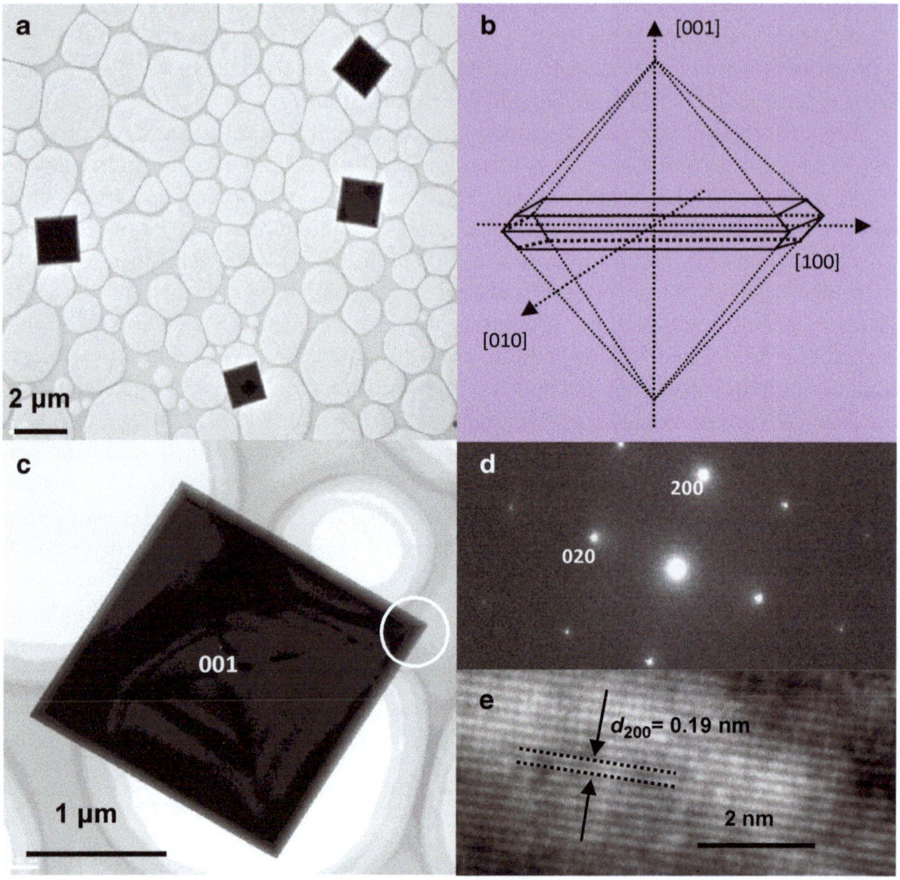

Fig. 4.3 (**a**) TEM images of anatase TiO_2 single crystal, (**b**) Schematic diagram of the crystal orientations in the anatase, (**c**) Magnified TEM image of the anatase single crystal, (**d**) and (**e**) SAED pattern and TEM image of the anatase crystal. (Reproduced from Yang et al. 2009)

where ρ (g cm^{-3}) is the density of the semiconductor samples, 6 is a factor encapsulated to serve spherical and cubic grains (used here not to mean agglomerates). Ardizzone et al. (2007) have shown that the average sizes of the crystallites as obtained from X-ray profile analysis (G_{XRD}), the size obtained by TEM (G_{TEM}), and the particle size obtained by elaboration of the specific surface area (G_{BET}) using Eq. (4.2), are fully comparable in the case of the sol-gel samples, and for the hydrothermal sample the agreement is excellent. Most non-porous substances would exhibit surface roughness (Rouquerol et al. 1999). Degussa P25 is not an exception. Arguably therefore, for such materials the crystallite size will be better estimated from both the surface area and XRD data, as perfectness of surface may not be ascertained. Yang et al. (2000) has reported both G_{BET} and G_{XRD} for hydrothermally synthesized nano-TiO_2 but in most cases the former nearly doubled.

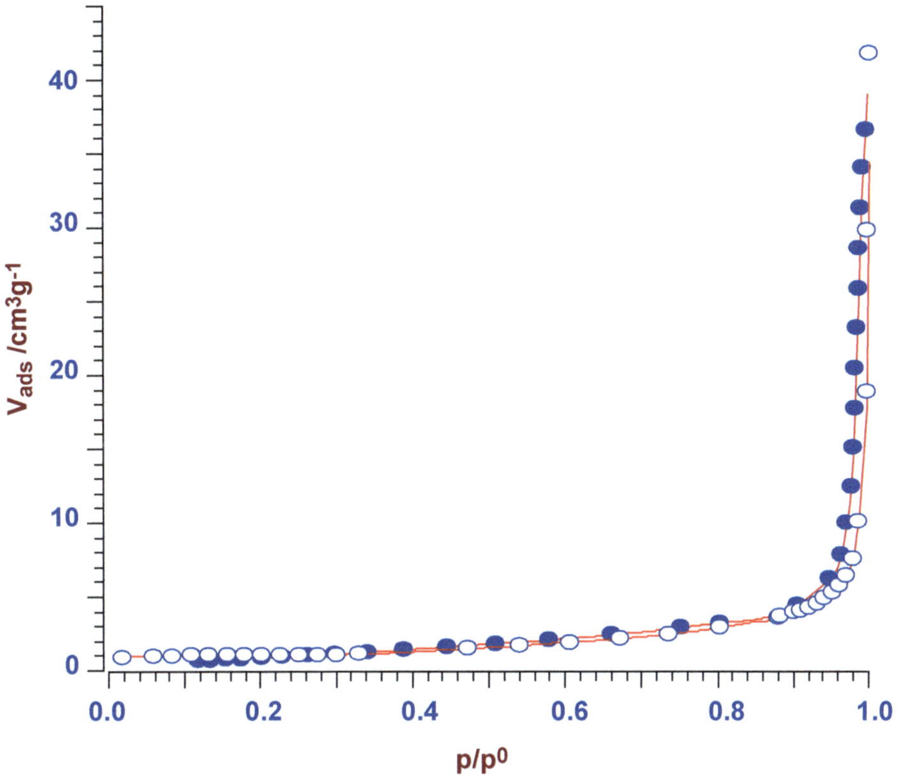

Fig. 4.4 Static BET adsorption-desorption isotherm of N_2 (gas volumes versus versus partial pressures) on non-porous ZnO (Merck), recorded on Thermo Finnigan Sorptomatic 1990 Series analyser, operated at 77 K

Apart from adsorption, photocatalytic performance can be influenced by diffusion behaviour of target molecules. In Chap. 3, we have discussed in detail the effect of diffusion of target molecules. Consequently, variations in macroscopic morphology of the photocatalysts can influence photocatalytic performance. Currently, hydrothermal methods have been relied upon to synthesise various architectures of highly crystalline semiconductor nano-solids with significant photoactivity under low-temperature (Wang et al. 2008). In a study using methylene blue, Kajbafvala et al. (2012) has found that the photocatalytic performance of ZnO nanostructures with spherical morphology and lower surface area is greater than that of flower-like structures. Similarly, the photocatalytic removal of the same dye on anatase TiO_2 preparations having the same crystal structure and band gap showed morphology dependence (Baiju et al. 2009). Surface morphology of photocatalysts may depend on several factors including the synthesis method, chemicals, and condition such as pH and temperature (Zhang et al. 2004; Jiang and Wang 2007). In fact, growth control chemicals may also be augmented to achieve desired results. For instance, by hydrothermal route Xi and Shang (2007) synthesized TiO_2 crystals of various morphologies: 1D nanowire structure from a base system of $Ti(C_3H_7O)_4$ and ethylene glycol, nanorods and fibres by using ethylene diammine as a growth inhibitor, and 2-D and 3-D nanostructures by increasing temperature.

4.2.2.2 Energy-Driven Processes

The primary energy driven process in photocatalysis is the band gap excitation. The band gap absorption edge of the semiconductor photocatalyst determines the region of light spectrum that should be used for electronic excitation. For example, TiO_2 with band gap of 3.2 eV will absorb light in the UV-A range wavelength, equal to or larger than 388 nm. Accordingly, component of sunlight available for absorption by titanium dioxide will be confined to less than 5 %. As electron is dislodged from the valence band it diffuses fast to the surface. There are chances of the extermination of this electron by hole in the semiconductor bulk, subsurface or surface by a process known as recombination. This process has been introduced in Chap. 1. The loss of charge carriers by these recombination processes is shown by Eq. (4.3).

$$e^-{}_{CB}/e^-{}_{TR'} + h^+{}_{TR'}/h^+{}_{VB} \rightarrow \text{energy(light/heat)} \tag{4.3}$$

where $e^-{}_{TR'}$ and $h^+{}_{TR'}$ are the shallowly-trapped electron and hole.

The presence of good electron acceptor such as oxygen would help lessen recombination as a result of which the quantum efficiency for photocatalytic oxidation then becomes high (Robertson et al. 2005). While the trapping of charge carriers is beneficial, recombination is a negative development by which hole concentration is reduced (Moser et al. 1987). However, both recombination and trapping processes are important in describing the efficiency of the photocatalytic oxidation (Shchukin et al. 2004; Neppolian et al. 2007).

4.2.3 Secondary Determinants of Photocatalytic Efficiency

4.2.3.1 Stability of the Photocatalyst Surface

Photocatalysis is an adsorption-based process and therefore it is morphology reliant. Therefore, catalyst surface ought to be photostable, photochemically resistant, photoactive and effective lest morphology gets altered and performance impaired. So far, TiO_2 has been the most superior photocatalyst combining these requirements. Cadmium sulphide with band gap edge of 2.4 eV can absorb most of the visible light in the solar spectrum and has great potential application in solar cells and photocatalysis, but the stability of CdS for a long period under photocatalytic condition was one of the most important hurdles against its use (Shen and Guo 2008). The preferential use of TiO_2 over ZnO in some applications such as in TOC analyser is mainly attached to relatively the incongruity of the ZnO under long and constant photochemical conditions. Little though it is, some dissolution occurs when ZnO is used as photocatalyst (Lathasree et al. 2004). This dissolution can either be acidic or alkaline depending on solution pH. The following reactions (Eq. 4.4 and 4.5) have been proposed to account for such pH dependent dissolutions (Zhang et al. 2009a).

In acidic medium:

$$ZnO+2H^{+} \rightarrow Zn^{2+} + H_2O \tag{4.4}$$

In basic medium:

$$ZnO+H_2O+2OH^{-} \rightarrow Zn\ [(OH_4)]^{2-} \tag{4.5}$$

Similarly, many workers have observed photocorrosion with long irradiation of ZnO (Maeda et al. 1982; Zhang et al. 2009b). Photocorrosion takes place by the following equation.

$$ZnO+2h^{+}_{VB} \rightarrow Zn^{2+} + {}^{1}\!/_{2}\,O_2 \tag{4.6}$$

The potential of modifying semiconductor photocatalysts with protective organic material to improve corrosion resistance while maintaining or even enhancing photocatalytic effect has been demonstrated. Vohra and Tanaka (2001) have shown enhancement in the photocatalytic degradation of aqueous paraquat (i.e. 1,1-dimethyl-4,4-bipyridinium dichloride) in presence Nafion (an anionic polymer)-coated TiO_2 and high photostability of this material against photocatalysis. Later on, Comparelli et al. (2005a) reported that protecting surfactant-capped photocatalyst may lead to higher efficiency of organic destruction than the uncapped catalyst despite the presence of surface passivating agents and hindering chains. These workers employed oleic acid- and tri-n-octylphosphine oxide- capped anatase titania nanocrystal powders to degrade model dyes. The destruction efficiency of oleic acid-capped anatase TiO_2 doubled that of both Degussa P-25 and tri-n-octylphosphine oxide-capped anatase titania. Similarly, it has also been shown by the same workers that surface-protected ZnO may outlast in the event of photocorrosion and pH changes. In the study, modified ZnO demonstrated higher performance over conventional ZnO photocatalyst (Comparelli et al. 2005b).

4.2.3.2 Acidity Dependence of the Surface Phenomena

The surface charge properties of the photocatalyst and the size of aggregates it forms in solution are largely controlled by solution acidity. To gain necessary information about the effect of pH on surface chemisorption an essential determining parameter is the point of zero charge (pzc) at which the surface of the semiconductor photocatalyst has no net charge. It is important to distinguish between point of zero charge and its analogue, the isoelectric point. While the former is determined by potentiometric methods the latter is measured by electrokinetic methods and can be comparatively lower for TiO_2 (Kallay et al. 1986). Therefore these terms can not be used interchangeably.

Basically, surface protonation or deprotonation of semiconductor photocatalysts may occur as a result of shifts in pH. In aqueous solutions, semiconductor oxides appear as metal with –OH functionality (M^+OH^-) so long as the pH remains in the near the point of zero charge of the photocatalyst. At this stage, progressively significant production of hydroxyl radical has been shown to occur, resulting in efficient photo-oxidation (Shourong et al. 1997; Gaya et al. 2009). When the pH of solution is varied away from the point of zero charge, acid-base dissociation equilibria may result from the protonation or deprotonation of the phenolic function, as shown in Eq. (4.7).

1. In acidic medium, the semiconductor surface is protonated and will release proton with increase in pH towards the pzc

$$MOH+H^+ \xrightarrow{pKa_1} MOH_2{}^+ \tag{4.7}$$

$$\text{e.g. Zn-OH+H}^+ \rightarrow ZnOH_2^+$$

2. With further increase in pH the semiconductor surface gets deprotonated by a second ionisation process

$$MOH+OH^- \xrightarrow{pKa_2} MO^- + H_2O \tag{4.8}$$

$$\text{e.g. Zn-OH+OH}^- \rightarrow ZnO^- + H_2O$$

Mathematically, pH_{pzc} is the average of the negative logarithm of the acid dissociation constant for the first ionisation Ka_1 (Eq. 4.7) and the second ionisation Ka_2 (Eq. 4.8) in the above processes. This parameter is represented mathematically by Eq. (4.9). Zinc oxide for example can have pH_{pzc} within the range of 8.6 to 9.8, depending upon preparation (Boxer 1997).

$$pH_{pzc} = \frac{(pKa_1 + pKa_2)}{2} \tag{4.9}$$

Phenolic compounds constitute a large group of pollutants, usually weak organic acids (Degen and Kosec 2003). Depending upon the pK_a of the pollutant, the dissociation of ionic substrates may be governed by solution pH. At low pH the phenolic-OH gets protonated. Conversely, as pH sparingly approaches the pK_a of the compound phenolate is formed by deprotonation. Excessively low or high pH would result in preclusion of adsorption since both the semiconductor photocatalysts and the phenolate would be electrostatically negative or positive, respectively. Using *o*-cresol as example, the changes in the phenolic charge can be given by the scheme below Eq. (4.10).

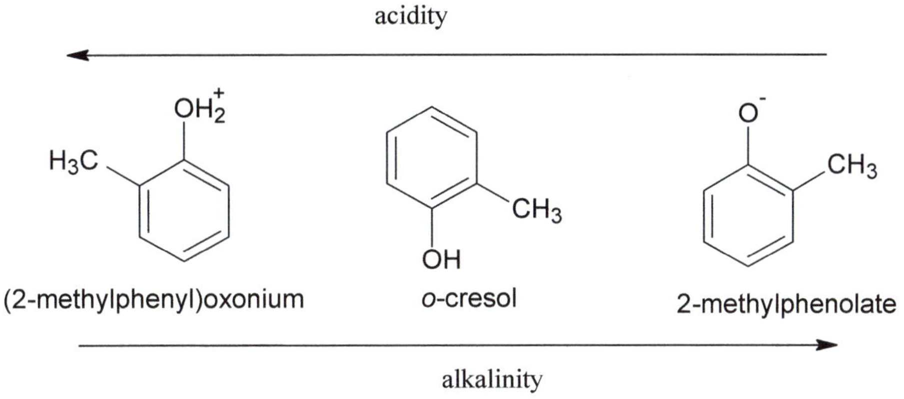

(4.10)

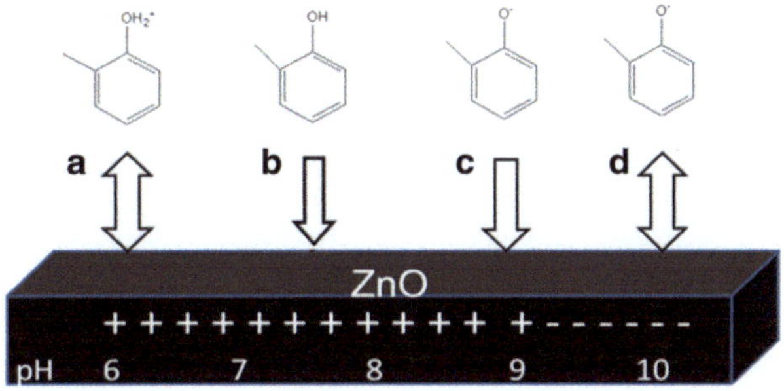

Fig. 4.5 Schematic diagram of ZnO surface and *o*-cresol in different solution pH. Several events are illustrated: (**a**) repulsion, (**b**) interaction, (**c**) effective attraction, (**d**) repulsion. The signs + and − indicate surface charge concentration on the ZnO semiconductor. (Reproduced from Abdollahi et al. 2011)

Recently, we have also illustrated the electrostatic effect resulting from pH changes during *o*-cresol destruction in presence of Mn-doped ZnO (Fig. 4.5).

The Fig. 4.5 shows the formation of methyl phenolate in the basic medium while methylphenyl oxonium is formed in acidic medium. It would be seen that exceedingly low or high pH leads to effective repulsion sufficient enough to deter the adsorption of *o*-cresol on the surface of semiconductor photocatalyst. At $pH > pH_{pzc}$ the surface of the semiconductor is negatively charged and the concentrations of any added anions will be lower nearer to the surface of the catalyst (Calza and Pelizzetti 2001). Speaking generally, electrostatically repelled substrates would diffuse into the bulk solution. In such a case, we can relate the concentration of monovalent anions (phenolates, Cl^- e.t.c.) on the surface of the semiconductor $[A]_s$ to the concentration in the bulk solution $[A]_b$ by Eq. (4.11) (Calza and Pelizzetti 2001; Slavchov et al. 2007).

$$[A^-]_s = \frac{\sigma_{pH}}{2\,\varepsilon\varepsilon_o KT} + 2\,[A^-]_b \tag{4.11}$$

Where σ_{pH} is the surface charge at a given pH, ε_o is the permittivity of vacuum, ε is the dielectric constant of the system, K is the Boltzmann constant and T is temperature. E is based on the dielectric system of the surface and the bulk solution.

4.3 Methods of Synthesis and Development of Semiconductor Catalysts

There is no gainsaying that interest in the synthesis of semiconductor compounds increases exponentially as their applications are discovered. As a result, variety of methods are constantly being developed for the synthesis of semiconductor solids. Some of the common methods exploited in the synthesis of semiconductor photocatalysts are explained in the following subsections.

4.3.1 Sol-Gel Method

Sol-gel is one of the most versatile routes to catalysts, particularly metal oxides. In the sol-gel preparation the following advantages can be harvested (Ward and Ko 1996). Single-component photocatalysts can achieve very high purity and textural properties. The surface area and crystallinity of these photocatalysts can be easily tailored. In multicomponent systems there is ability to control structure and composition at molecular level, to introduce several components in a single step, to impose kinetic constraints on a system thereby stabilising metastable phases, and the ability to fine tune the activation behaviour of a sample or consequently trace the genesis of active species.

Sol-gel a name given to a process in which dispersions of a colloidal suspension known as the sol is formed in a liquid by the hydrolysis of inorganic or metal organic precursors, which is then followed by gel formation. The commonest group of precursors used in sol-gel preparations is the metal alkoxide, $M(OR)_x$ where R is an alkyl group and M is the metal and x=1 to 5 representing the number of alkoxy groups. Generally, alkoxides are fairly strong bases, known to react rapidly with a wide variety of chemical compounds (Clayden et al. 2001). Basically, the partial hydrolysis of metal alkoxides is a rapid step leading to the formation of intermediate $M(OR)_{n-z}(OH)_z$ which respond as dispersions of colloidal sol in different regions. Sol undergoes condensation reaction with the starting alkoxide or self to cooperatively yield M-O-M linkages that coalesce as infinite network of submicrometer gel particles. Water or alkanol is invariably loss in the course of the reaction. We can simply represent the under laying chemical steps of sol-gel transition by the following equations.

Hydrolysis:

$$M(OR)_x + H_2O \rightarrow M(OR)_{x-z}(OH)_z + zROH \qquad (4.12)$$

e.g. $$Ti(\textit{tert}\text{-}OC_4H_9)_4 + H_2O \rightarrow Ti(OC_4H_9)_3OH{+}CH_3(CH_2)_2CH_2OH$$

Self-condensation:

$$M(OR)_{x-z}(OH)_z{+}M(OR)_{x-z}(OH)_z \rightarrow (OR)_{x-z}M\text{-}O\text{-}M(OR)_{x-z} + zH_2O \qquad (4.13)$$

e.g. $$Ti(OC_4H_9)_3OH + Ti(OC_4H_9)_3OH \rightarrow (OC_4H_9)_3Ti\text{-}O\text{-}Ti(OC_4H_9)_3 + H_2O$$

Mixed-condensation:

$$M(OR)_x + M(OR)_{x-z}(OH)_z \rightarrow (OR)_xM\text{-}O\text{-}M(OR)_{x-z} + zROH \tag{4.14}$$

e.g. $$Ti(OC_4H_9)_4 + Ti(OC_4H_9)_3OH \rightarrow (OC_4H_9)_3Ti\text{-}O\text{-}Ti(OC_4H_9)_3 + CH_3(CH_2)_2CH_2OH$$

Hydrolysis process may go to completion in presence of sufficient water (Su et al. 2004) and efficient agitation, so that all alkoxy groups are replaced by hydroxyl groups.

Completion of hydrolysis:

$$M(OR)_n + H_2O \rightarrow M(OR)(OH)_n + nROH \tag{4.15}$$

e.g. $$Ti(OC_4H_9)_4 + H_2O \rightarrow Ti(OH)_4 + 4CH_3(CH_2)_2CH_2OH$$

Gel needs to be processed to remove solvent, cause crystallisation and nanosizing. Besides, some sol is still present within the gel with which activities such as condensation and phase changes may proceed. The process of change in structure and properties after gelation is called aging (Brinker and Scerrer 1990). Aging is usually carried out at room temperature but can be aided by long solvothermal treatment at 80–100 °C. A sufficiently aged gel undergoes further polycondensation, reduces surface area without modifying pore size and coarsens to resists cracking during calcinations (Hench and West 1990). In a subsequent step, the pore liquid within the gel can be dried by evaporationbelow 180 °C to form xerogel (if water is removed) or alcogel (if alcohol is removed). This form of drying proceeds under capillary pressure. In porous semiconductor photocatalysts such the capillary pressure increases inversely proportional to pore size, according to modified Young-Laplace's equation expressed below.

$$\Delta P = \frac{2\gamma(\cos\theta)}{r_p} \tag{4.16}$$

Where ΔP is the pressure difference in the capillary, γ is the surface tension at the interface, θ is the contact or wetting angle and r_p is the pore radius. The dried gel is calcined typically at 350–450 °C for few hours to remove chemisorbed hydroxyls, organic residues and finally produce the desired catalyst. For successful synthesis of nano- or hollow TiO_2, the gel may be subjected to prolonged heat treatment below 200 °C, typically half day to a few days (Zhang et al. 1998; Sudhagar et al. 2008).

Some economic disadvantages of the use of alkoxides in material synthesis include high cost and the need to handle organic solvent. Inorganic salts such as $TiCl_4$ have therefore gained preference as precursors. Gel can be derived from the hydrolysis of $TiCl_4$ to $Ti(OH)_4$. Acids such as HCl and HNO_3 are the commonly used pH adjustment solutions. Most often, the hydrolysis is carried out at low pH so as to control the particle size of the catalyst. Su et al. (2004) have observed that around the point of zero charge, alkoxide derived sol may become milky-white, accompanied by precipitation which indicates the rapid coagulation of TiO_2 particles (as suggested by the pH dependent surface potential). Nevertheless, clear sol with nanosized TiO_2 can be obtained at pH$<$3 (which can be maintained for example by HNO_3 addition) and/or pH$>$9. Therefore, a suitable pH is required for successful catalyst synthesis by the sol-gel method. Due to these reasons, many workers have preferred to carry out the synthesis of TiO_2 nano-particles in presence of HCl, rather than in neat deionised water (Wei et al. 2004).

To immobilise uniform catalyst film on substrate by sol-gel method, dip-coating or spin-coating technique is employed. Automated coaters are commercially available, to keep operator error minimal and accurately maintain dipping speed of substrate and duration. The film thickness can be correlated with the angular velocity, the solvent evaporation rate and the viscosity of the fluid by Meyerhofer equation (Augugliaro et al. 2010):

$$H = \left(1 - \frac{\rho_{\mathrm{A}}}{\rho_{\mathrm{A}_o}}\right)\left(\frac{3\mu m}{2\rho \mathrm{A_o}\omega^2}\right)^{1/3} \tag{4.17}$$

where H is the final thickness, ρ_A and ρ_{Ao} are the initial and final density of the volatile solvent respectively, μ is the viscosity, m is the rate of evaporation of the solvent and ω is the angular speed. This equation can be encapsulated into a simpler form:

$$H = \mathrm{A}\omega^B \tag{4.18}$$

The equation is similar to Eq. (4.17) as the terms A and B must also be experimentally determined from rheological studies.

Sol-gel route can be used to synthesise coupled as well as doped semiconductor photocatalysts. Liao et al. (2004) have prepared ZnO-TiO_2 from ethanolic n-butyl titanate, citric acid, precursor of zinc, water and polyglycol at room temperature. The mixture was stirred for an hour to obtain a yellow sol, which was aged for 24 h at room temperature, dried at 80 °C for 48 h and pre-treated at 200 °C for 1 h. The obtained dry gel powder was ground and followed by calcining at various temperatures. A series of binary oxide catalyst, in which the zinc concentration is different, was prepared which showed substantial improvement of the activity of TiO_2 for methyl orange degradation.

4.3.2 *Hydrothermal/Solvothermal Synthesis*

Hydrothermal synthesis method is based on the reaction of aqueous precursors in a sealed pressure vessel which is usually a bare or Teflon-lined stainless steel autoclave at above ambient, autogenous pressure and mild temperature (typically $\leq 220\,°C$). This low-temperature technique is critical to the formation of metastable phases that are not obtainable using traditional high-temperature methods (Whittingham 1996). On the other hand, solvothermal route differs only by the fact that precursors are non-aqueous. The use of non-aqueous solvent in the solvothermal route prevents the oxidation of the products, which is critical in the synthesis of non-oxide materials (Zou et al. 2006). Even though separation of catalyst may be difficult, the addition of surfactants such as ethylene glycol and poly (sodium-p-styrene sulfonate) is among common practices in solvothermal methods (Shao et al. 2005). Owing to these flexibilities, solvothermal method offers better control of the size, morphology and crystallinity of nano-scale materials.

Solvothermal method has been used in the synthesis of the exceptionally fascinating, single atom-thick graphene. Differently from hydrothermal route, temperature is occasionally raised much higher in the solvothermal method. Solvothermal route can tailor properties to give potentially useful catalysts for different applications including for water splitting. For example, brookite titania with improved crystallinity and sufficient surface area was obtained by Kominami et al. (2003) by the solvothermal treatment of oxobis(2,4-pentanedionato-O, O′)titanium ($TiO(acac)_2$) in ethylene glycol (EG) in the presence of sodium acetate and a small amount of water at 300 °C. The synthesis was carried out in a water-bathed test-tube, placed in autoclave. The product was calcined at 700 °C yielding brookite exhibiting hydrogen evolution rate almost equal to P25. Similarly, ZnS with high photosplitting ability has been synthesized solvothermally from autoclaved mixture of $Zn(NO_3)_2 \cdot 6H_2O$ and thiourea NH_2CSNH_2 in excess of ethylenediamine $NH_2(CH_2)_2NH_2$ at 160 °C for 12 h (Ding et al. 2001). The precipitate obtained was washed with ethanol and deionised water. The ZnS photocatalyst was obtained after treatment of the precipitate at 80 °C for 12 h and calcination at 500 °C. Generally, the important point to note is that there is no sharp distinction between hydrothermal and solvothermal method. In actual fact, some authors appear to use the terms hydrothermal and solvothermal equivalently or with aqueous/non-aqueous appended to the term solvothermal. More importantly to note, both methods have systematised the construction of nanoscopic semiconductor and non-semiconductor materials in different shapes as nanorods, nanospindles, nanosheets and so on, which cannot be derived by traditional methods.

There is a wide range of precursors for hydrothermal route from metals, inorganic salts to metal-organics (Jang et al. 2008). The hydrothermal process can proceed without any templates or additives though can be assisted by the addition of mineral acids (Yu et al. 2007). The addition of acid improves dispersion or in some cases, solubility. It has been observed that acid that coordinates strongly to titanium would behave as a poor additive. In a study, Wu et al. (2002) prepared some uniform and unaggregated nanocrystals of pure anatase and rutile from various acidic mediums. Their findings revealed that nanocrystals of pure rutile with sizes <10 nm can be obtained at higher HCl concentrations under much milder hydrothermal conditions, such as 140 °C. The ability of an acidic medium favouring the formation of rutile follows the order $HCl > HNO_3 > H_2SO_4 > CH_3COOH$, which appears to be unrelated with the pKa of the acids. Sodium chloride (NaCl) is as well a good additive, but may physically adsorb on the TiO_2, thereby confining crystal growth. Similarly, Yang et al. (2009) synthesized nanorutile TiO_2 particles of 100 % phase purity and high specific surface areas hydrothermally at a relatively low temperatures, by peptizing sols derived from $Ti(OC_3H_7)_4$ and $Ti(OC_4H_9)_4$ in presence of HNO_3. Particles that contained highly crystalline nanoanatase, together with a small portion of rutile, were formed in both tetraethylammonium hydroxide-peptized samples. The same workers prepared high-quality anatase TiO_2 single-crystal nanosheets mainly dominated by (001) crystallographic face using a water-2-propanol hydrothermal synthetic route. The precatalyst TiF_4 was mixed with HF in a Teflon-lined autoclave kept at 180 °C for 5.5–44 h in an electric oven. The anatase TiO_2 single-crystal had 64 % (001) facets (Fig. 4.6a) and average size of 1.06 μm (Fig. 4.6b and c), and showed superior photoreactivity (more than 5 times) compared to the benchmark P25.

The idea of varying volume ratio of water to alcohol in solvent was employed by Lei and Duan (2008) to yield phase composition-controllable TiO_2 photocatalysts via a low temperature solvothermal route at 80 °C. It was found that anatase contents increase from 0–100 % with increase of ethanol contents in the solvent while the enhancement of photocatalytic activity followed the order 100 %<0 %, 5 %<10 %<60 %. This order also reinforces the expectation of relatively poor activity of 100 % anatase or 100 % rutile titania. In actual fact, a low H_2O/Ti molar ratio was necessary for preparing anatase-rutile composites, because the formation of photocatalytically inactive brookite phase was favored at relatively high H_2O/Ti molar ratios, but an optimum HCl/Ti molar ratio is critical for rutile formation (Liu and Zeng 2003).

Zinc oxide can be prepared hydrothermally from its soluble salts. Apart from acids or bases (NaOH, HNO_3 or HCl), a wide variety of chemical additives including alcohols (such as ethanol, butanol, isopropanol), ammines (such as ethylenediammine, hexamethylenetetramine), sodium salts (such as sodium chloride, sodium citrate) have been added together to fabricate different nanoproducts (Li and Gray 2007). Kuo et al. (2005) synthesized ZnO substructures from dilute aqueous solutions of $Zn(NO_3)_2 \cdot 6H_2O$, hexamethylenetetramine, and trisodium citrate in a Teflon-lined autoclave fixed at 95 °C for 30 min to 48 h. The resulting white precipitate was collected and washed with deionized water several times, dried in an oven at 65 °C and then tested for the photocatalytic degradation of phenol to obtain nanosheets. These nanosheet structures did not display photocatalytic activity due to possible surface adsorption of solution species. However, when the microstructures were activated by heat treatment to 300 °C there was substantial enhancement in the photodecomposition of phenol under direct sunlight irradiation while still maintaining high surface area.

Nowadays, the synthesis of hollow spheres instead of massive nanoparticle spheres is gaining much interest. Yu and Yu (2008) have demonstrated the synthesis of ZnO hollow spheres with porous crystalline shells by hydrothermal treatment of glucose/$ZnCl_2$ mixtures at 180 °C for 24 h, and calcinations at 500 °C for 4 h.

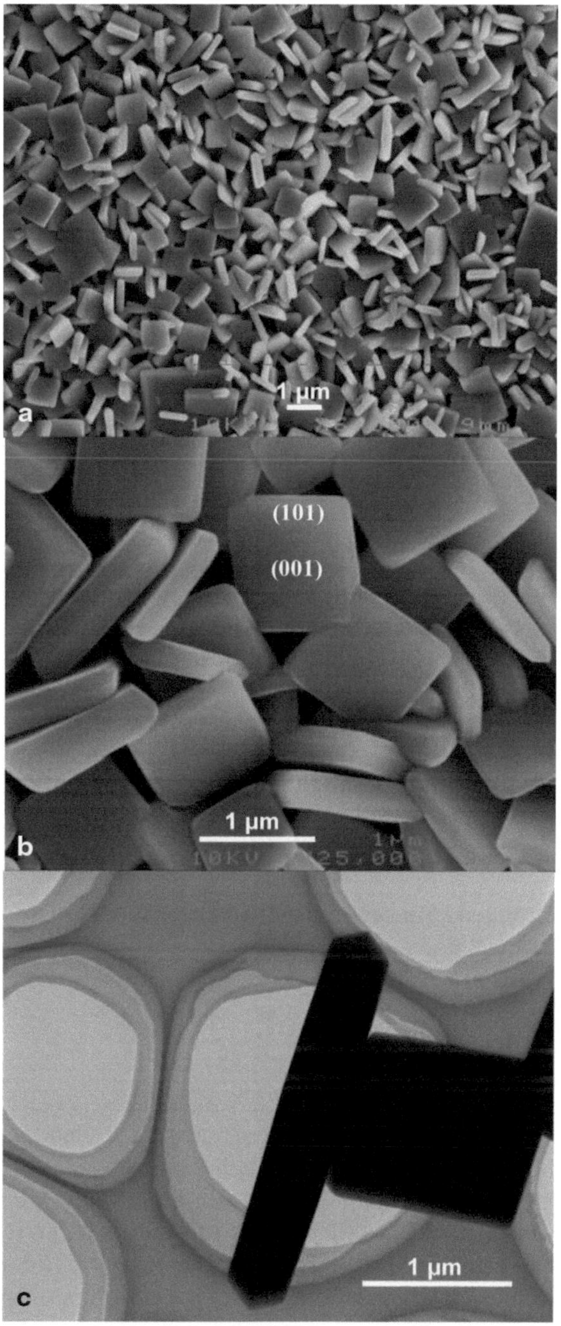

Fig. 4.6 Images of hydrothermally produced anatase TiO_2 single-crystal nanosheets recorded on SEM (**a**) and TEM (**b** and **c**). (Reproduced from Yang et al. 2009)

Their findings indicated that the average crystallite size, shell thickness, specific surface areas, pore structures, and photocatalytic activity of ZnO hollow spheres could be controlled by varying the molar ratio of glucose to zinc ions. Hydrothermal/solvothermal method can be used to produce composite semiconductor photocatalysts. Composite of ZnO-ZnO_2 has been prepared hydrothermally by subjecting ZnO_2 to hydrothermal treatment at 120 and 180 °C for 2 h (Hsu and Wu 2005). The ZnO_2 was obtained from a precipitate formed by mixing aqueous $ZnSO_4 \cdot 7H_2O$ and NH_4OH. The precipitate was washed with excessive amount of deionized water and then introduced into H_2O_2. The resulting dispersion was then heated at 75 °C for 2 h to form ZnO_2 powder, which was subsequently washed with excessive deionized water and finally dried at 65 °C in air.

In hydrothermal methods, surfactant can be used to provide desirable dimensional nanostructures that may require effort to harness by other methods. Surfactant-assisted method of synthesis is usually referred to as the 'polyol' method. In polyol-mediated routes the precursor is refluxed at mild temperature (<200 °C) in presence of a suitable surfactant. A polyol method based on reduction of chloroplatinic acid at 170 °C has been used to produce pure platinum nanosizes for loading onto TiO_2 (Sivalingam et al. 2003). In a different study, indium and nitrogen co-doped TiO_2 has been prepared for photocatalytic hydrogen production using polyol method based on refluxing a mixture of Ti, $InCl_2$, HCl and ethylene glycol at 100 °C for 1 h (Sasikala et al. 2010). Similarly, Li et al. (2003) have demonstrated the polyethylene glycol assisted hydrothermal synthesis of ZnO nanowires and nanorods at 140 °C (held for 24 h), using NaOH treated $Zn(CH_3COO)_2$ solution. Short chain polymer (PEG400) was found to promote the formation of 1D ZnO nanostructures, which cannot be obtained by long-chain polymers (such as PEG10000).

4.3.3 Co-Precipitation Methods

Ordinarily, the synthesis of semiconductor particles from solution using the separate precipitation method involves the precipitation of metal salts usually in basic solutions of LiOH, NH_4OH or NaOH. For instance, Zn^{2+} can be derived from some many common salts which include $ZnCl_2$, $Zn(NO_3)_2$, $ZnSO_4$, $Zn(CH_3COO)_2 \cdot 2H_2O$. As a result, precipitation has been one of the commonest methods for the preparation of bulk ZnO photocatalyst. A comparative study of different combinations of precipitating solution and precursor source, by reaction at pH 7 and drying precipitate at 110 °C puts $Zn(OOCCH_3)_2$ and $NaHCO_3$ solution at the forefront, as the ZnO produced showed the highest efficiency in phenol degradation (Parida and Parija 2006). The precursor sources examined in the study were (i) $Zn(CH_3COO)_2$ and NaOH, (ii) $ZnCl_2$ and NaOH (iii) $Zn(NO_3)_2$ and Na_2CO_3 (iv) $Zn(NO_3)_2$ and NH_4OH. If desired, different precursor sources can be made to co-precipitate as non-stoichiometric semiconductor catalysts, cocktail of mixed or doped catalysts. In addition, the co-precipition of two or more substances can occur directly without the addition of pH adjustment chemicals such as alkaline solutions. For example, Fe^{3+} -doped anatase TiO_2 has been prepared by simple co-precipitation with aqueous $TiCl_4$ (Ranjit and Viswanathan 1997). Homogeneous co-precipitation offers enormous benefits over

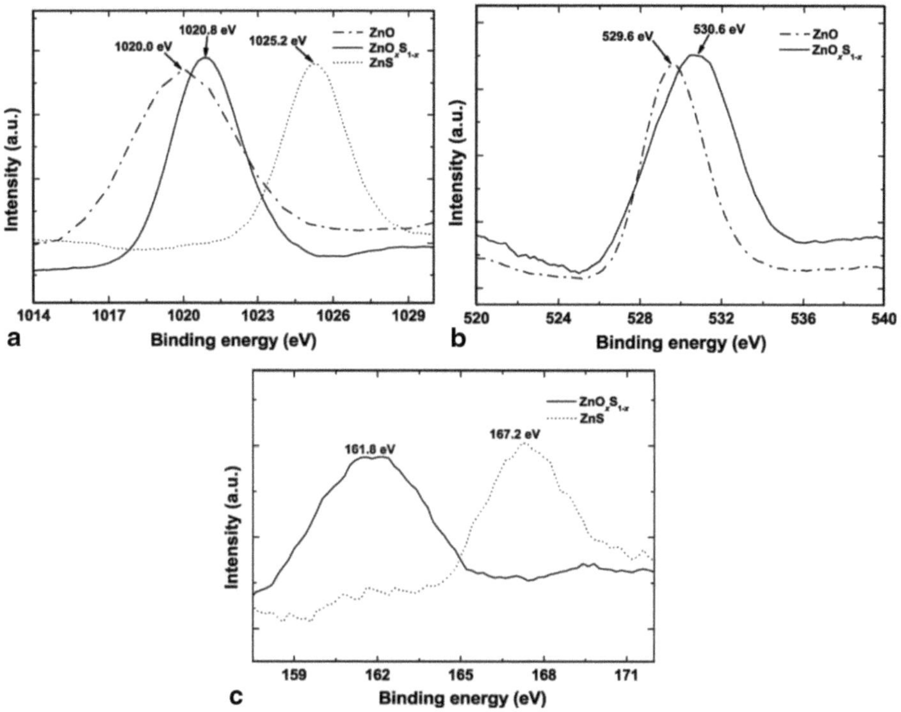

Fig. 4.7 X-ray photoelectron spectra of ZnO, ZnS and their composite photocatalyst, ZnO_xS_{1-x} prepared by co-precipitation. (Reproduced from Kim et al. 2007)

other methods of synthesis ranging from quantitative precipitation, inexpensiveness, varied choice of starting materials to simple methods of removal of co-precipitated impurities.

Co-precipitation is a convenient method for the introduction of dopant. In order to gain a visible light-driven composite semiconductor photocatalyst, Kim et al. (2007) prepared ZnO_xS_{1-x} from aqueous solution of $Zn(NO_3)_2$ and Na_2S by precipitation using NaOH. The catalyst substantially removed 4-chlorocatechol and basic red 2 from aqueous solution. X-ray diffraction (XRD) and energy dispersive X-ray (EDX) analyses indicated that the obtained crystals are not the simple mixtures of ZnS and ZnO but their solid solutions. Zinc was confirmed to be chemically bonded to sulphide and oxide at the same time by X-ray photoelectron spectroscopy (XPS) as shown in Fig. 4.7, which was reinforced by findings based on DRS. Inspection of the XPS spectra shows the Zn $2p_{3/2}$ binding energies for ZnO and ZnO_xS_{1-x} to be 1020.0 and 1020.8 eV (Fig. 4.7a) while the O1s binding energies for these semiconductors were 529.6 and 530.6 eV (Fig. 4.7b), respectively. On the other hand, the peaks of Zn $2p_{3/2}$ and S 2p binding energies for ZnO_xS_{1-x} were located at 1020.8 and 161.8 eV, respectively, accompanied by an absorption edge shift to lower energy compared to ZnS (Fig. 4.7a and c).

Table 4.1 Representative precipitation-derived semiconductor photocatalysts, summary of synthesis procedure and authors commenton their photoactivity

Catalyst and reference	Starting materials and preparation conditions	Authors remark
Mn-doped nano-ZnO (Ullah and Dutta 2008)	$Zn(CH_3COO)_2$ and $Mn(CH_3COO)_2$, were co-precipitated by NaOH, all alcoholised at 50 °C. Precipitate was derived by cooling mixtures in water bath	Photocatalyst is effective in methylene blue degradation
Ag-doped ZnO (Wang et al. 2004)	Ag^+ -sprayed $Zn(NO_3)_2 \cdot 6H_2O$ was stirred, adjusted to pH 7, heated to 60 °C and precipitated using NH_4HCO_3	Photocatalyst showed enhanced photoactivity for phenol and methylene blue degradation
Co-doped ZnO (Xiao et al. 2007)	$Zn(CH_3COO)_2$ and $Co(CH_3COO)_2$, were hydrothermally co-precipitated by KOH at 240 °C for 16 h. The product was collected by filtration, repeatedly washed with deionised water, then with ethanol and dried at 60 °C for 12 h	Co-doped ZnO photocatalysts showed substantial photocatalytic activity for methylene blue decolorization at pH 10.5 under visible light
Nanosized ZnO particles (20–50 nm) (Škapin et al. 2007)	Aqueous Na_2CO_3 was added to ethanolic $Zn(CH_3COO)_2$ to form hydrozincite. Nano-ZnO particles began to form on heating to 200 °C	At 600 °C, 20–50 nm particles were formed without agglomeration
La-doped ZnO (Anandan et al. 2007)	$La(NO_3)_3 \cdot 6H_2O$ and Na_2CO_3 solutions were added to $Zn(NO_3)_2 \cdot 6H_2O$ which has been under vigorous stirring. with distilled water, and then washed twice with ethanol. The resultant solid product was dried at 100 °C for 12 h and calcined at 300 °C for 2 h	The catalyst showed enhanced photocatalytic degradation of monocrotophos and trichlorophenol

Homogeneous precipitation process and resulting crystallinity of product can be favoured by controlling acidity, temperature, agitation and the use of dilute precursor solutions. Without agitation, heating and cooling precipitation can be indefinitely delayed. Agitation can be carried out by vigorous stirring or by ultrasonication (Bandara et al. 2002). One problem identified with co-precipitation is the association of the products with impurities. Often the approach for the removal of co-precipitated impurities from the precipitate is washing using suitable solvent. The extent of impurity removal can be expressed by simple qualitative testing. For example, photocatalytically active ZnO-SnO_2 nanoparticles have been prepared by a different procedure using 1:2 molar ratio of $SnCl_4 \cdot 5H_2O$ and $ZnSO_4 \cdot 7H_2O$ as starting materials and NaOH to adjust pH to 7 (Cun et al. 2002). To remove traces of SO_4^{2-} and Cl^- the precipitate was filtered and washed with deionized water till the disappearace of the ions in the filtrate. This process has often been followed by washing with ethanol (Table 4.1).

Although heat is used during precipitation, certain crystalline semiconductor photocatalysts can be synthesised at room temperature. Nanosized coupled oxide $ZnO–SnO_2$ was readily prepared by Zhang et al. (2005) using homogeneous co-precipitation method. The coupled oxide prepared had a better photocatalytic activity than that of either single oxides, ZnO or SnO_2, and its photocatalytic activity was slightly higher than commercial $P25 \cdot TiO_2$. Aqueous solution of NaOH was added dropwise into dilute $Zn(NO_3)_2$ under vigorous stirring until transparency to form $Na_2[Zn(OH)_4]$, and then aqueous Na_2SnO_3 was added. The precipitate was obtained at room temperature following the addition of 3:3:2 volume ratio of H_2O, CH_3COCH_3 and $CH_3COOC_2H_5$ until a pH 9–10. It was then washed 3 times with distilled water, dried at 80 °C, then calcined at different temperatures in air to yield $ZnO–SnO_2$ particles.

4.3.4 Vapour Deposition

TiO_2 exhibits a great deal of extraordinary properties such as anti-bacterial, cleaning and deodorization effects, which opened new page in its exceptionally wide industrial relevance. Thin films of TiO_2 are used in electrical applications (such as capacitors) and photocatalytic applications (such as photovoltaic cells). Ordinarily, TiO_2 thin films can be prepared by wet technologies such as sol-gel dip-coating or spin coating, liquid phase deposition, for various other applications such as self-cleaning glasses and tiles. When controlled structures with low defect density are desired to be produced, vapour deposition is a suitable method for such purpose. In addition, vapour deposition technique has found widespread acceptance in the production high quality semiconductor materials for various technical applications (Helmersson et al. 2006). There are two main vapour deposition methods: chemical vapour deposition (CVD) and physical vapour deposition (PVD). In PVD method, deposition occurs directly without chemical transformation of the precursor. However, the PVD method entails difficulty in the production of large area TiO_2 films on glass which makes it unsuitable for the production of self-cleaning glass (Sun et al. 2008). Chemical vapour deposition is much more relevant for practical applications owing to its flexibility as regards the shape of the substrate, including the capability to create films on the inner surface of pipes and on flexible substrates; compatibility with lithographic methods and glass fabrication processes; high quality of the prepared thin films and their good adhesion (Bessergenev et al. 2006). Figure 4.8 shows a schematic of chemical vapour deposition. Basically in CVD, precursors are heated at moderate temperatures and delivered to the reactor by carrier gas such as N_2, He or Ar. Reactive gases can also be supplied if desired. The substrates to be coated are deposited in the CVD reactor which usually operates at high temperature. As chemical transformation of the precursors takes place within the reactor, the semiconductor film is formed. Unlike in chemical vapour synthesis the coating of semiconductor films occurs via heterogeneous nucleation.

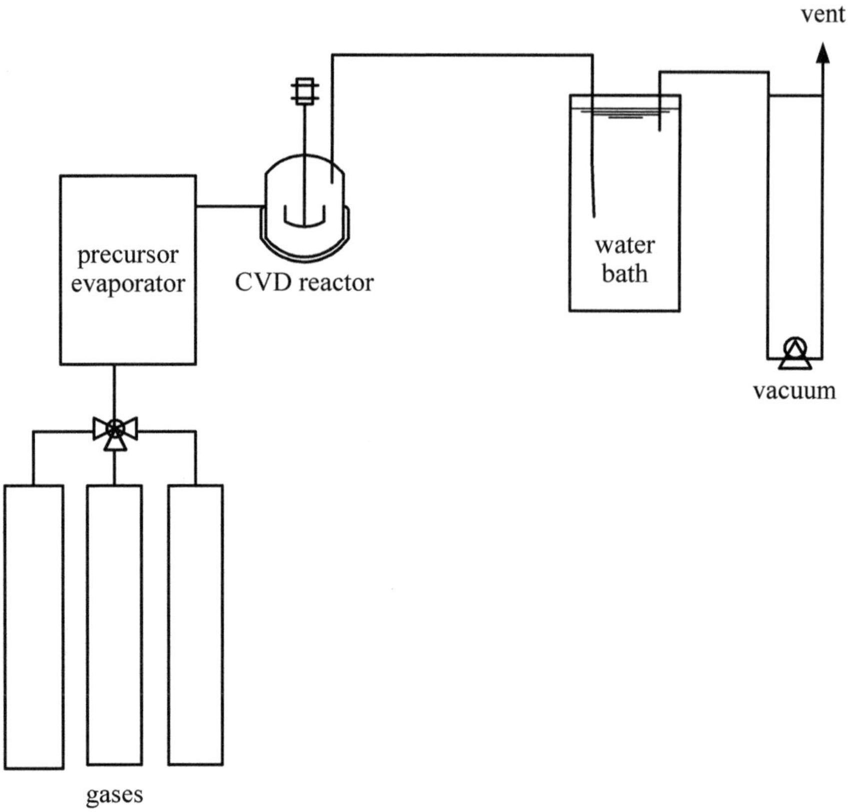

Fig. 4.8 A schematic representation of chemical vapour deposition

Basically, CVD methods differ by the activation approach, choice of precursor, temperature and pressure conditions for the reaction. Mostly, TiO_2 CVD films are deposited by thermal low pressure method (LPCVD). For example, the self-cleaning window glass, Activ^{TM} from Pilkington Glass comprised of a 20 nm coat of titania deposited by an atmospheric pressure CVD (APCVD) method (Mills et al. 2002). Titanium tetra-isopropoxide (TTIP) is preferred as precursor because it is relatively easier to evaporate and milder than $TiCl_4$ in reacting with water vapour (Ding et al. 2000; Lokhande et al. 2004).

TiO_2 is often produced by the pyrolysis of titanium chloride or the low boiling point titanium alkoxides such as titanium ethoxide and titanium isopropoxide as shown by Eq. (4.19) (Pierson 1999). A common example of this CVD is the oxidation of $TiCl_4$ at 400–1,000 °C.

$$TiCl_4 + O_2 + 2H_2 \rightarrow TiO_2 + 4HCl \tag{4.19}$$

Recently, two important chemical vapour deposition methods have rapidly progressed: the plasma-enhanced CVD (PECVD) and the metallo-organic CVD

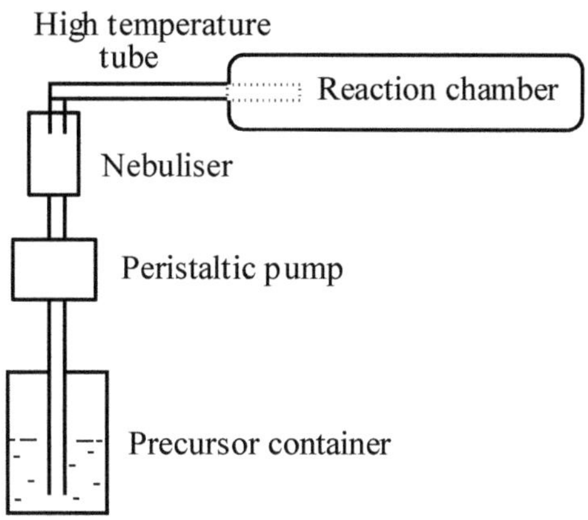

Fig. 4.9 Schematic diagram of spray pyrolysis set-up for the preparation of semiconductor coatings

(MOCVD). In MOCVD, $Ti(OC_2H_5)_4$ is made to undergo pyrolysis in an oxygen and helium atmosphere at 450 °C. Similarly, TiO_2 can be derived from the reaction of titanium tetraisopropoxide, $Ti(OC_3H_7)_4$, with oxygen at 300 °C and at low pressure (133 Pa). In spray pyrolysis, the precursors are pumped in the form of aerosol dispersions. The aerosol droplets may be passed through heated tube set at desired temperature and then into the reactor tube where the final powdered products are deposited (Hsiao et al. 2007). If a film of the semiconductor photocatalyst is desired to be produced, the substrate may be heated as the aerosol droplets are introduced (Diaz et al. 2007). A simplified diagram of a spray pyrolysis set up is shown in Fig. 4.9. Several variables may affect the decomposition of the precursors and consequently the quality of the semiconductor photocatalyst produced by spray pyrolysis process, which include the working temperature gradient and pressure of the carrier gas.

In the plasma enhanced CVD method, different TiO_2 with the same composition and binding states can be obtained, though properties like crystal structure, surface morphology and optical properties may vary (Huang and Chem 2002). Ayllón et al. (1999) prepared, by plasma-enhanced CVD (PECVD) and within short synthesis time, well crystallised TiO_2 particles that maintain a relatively high surface area after treatment at high temperatures, having good photocatalytic activity in the aniline photodegradation. The PECVD reactor was operated at external radiofrequency (RF) of 40 kHz, $Ti(OC_3H_7)_4$ was used as the precursor and oxygen as the carrier gas. The powder was deposited on the cold areas of the reactor at temperatures not higher than 90 °C. The titania powder was formed by vapour phase decomposition of the plasma-activated $Ti(OC_3H_7)_4$ precursor in an oxygen atmosphere. By PECVD, doped TiO_2 films can be synthesized (Randeniya et al. 2007).

4.3.5 Microemulsion Method

Microemulsion is a thermodynamically stable homogeneous dispersion of two immiscible liquids in the presence of an emulsifier or surfactant. The commonest types

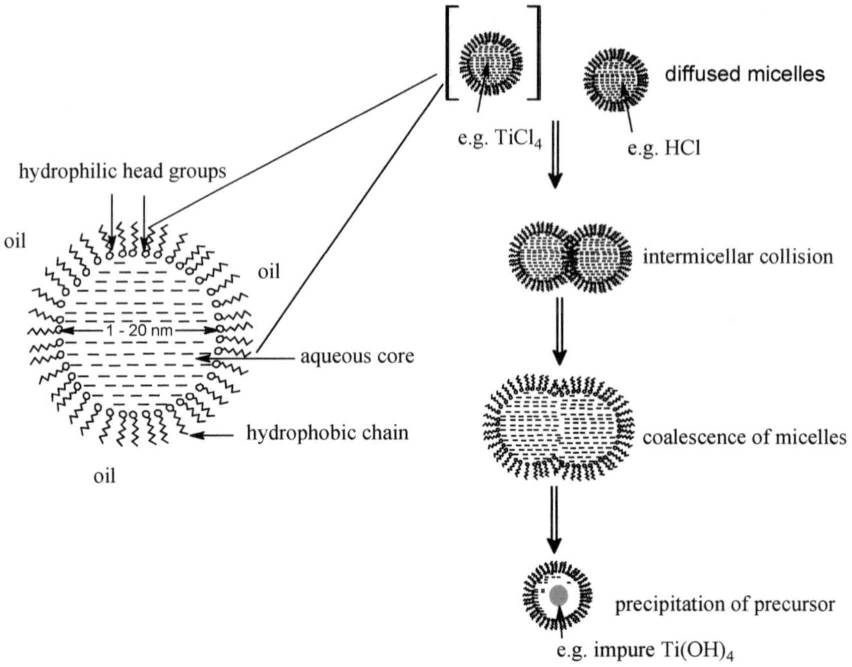

Fig. 4.10 Schematic diagram of reverse micelles and the processes leading to precipitation of the catalyst precursor. Shown on the *left* is the expanded structure of an inverse micelle unit residing in oil/water interface

of microemulsions are direct, when oil is homogeneously dispersed in water (o/w) and reversed, when water is homogeneously dispersed in oil (w/o). Basically, surfactants aggregate in various micellar structural forms at the oil/water interface. The normal micelles aggregate in aqueous phase while the reverse (or inverse) micelles organise themselves in non-polar organic solvents, characterised by water-surfactant ratio of <15. The latter are essential in the synthesis of subsizes of semiconductor particles. In microemulsion-based synthesis, inverse micelles act as microreactors containing the reactants which collide and coalesce as bigger droplets continuously to form products. The main advantage of microemulsions as spatially confined microreactors is their easy, one-step fabrication without sophisticated, time-consuming synthesis stages (Ganguli et al. 2010).

Microemulsion method is basically a surfactant-modified hydrothermal route. The reacting components are usually in ternary form consisting of surfactant, aqueous phase, co-surfactant and oil phase. Water is used as polar phase and oils such as cyclohexane and n-dodecane are used as non-polar phase. In normal o/w-microemulsion synthesis, the first reactant is added to the non-polar phase while the second reactant is added to the aqueous phase. Thus, in the synthesis of TiO_2, acids and titanium dioxide precursors are added to the aqueous phase. The alcohol n-hexanol is commonly used as co-surfactant while cetyltrimethylammonium bromide (CTAB) and Triton-X are used as surfactants. In the controlled synthesis of semiconductor nanoparticles w/o microemulsions (or reverse micelles) have been preferred for use as microreactors. The reactivity system is usually heated at mild temperature (50–200 °C) under constant stirring. The micellar microreactors collide with one another and exchange the reactants as shown in Fig. 4.10. Finally, the precursor is isolated from the microemulsions as precipitate using precipitating agents. In the preparation of semiconductor oxides, aqueous ammonia can be used to precipitate the hydroxide.

Several workers have synthesised semiconductor photocatalysts using the microemulsion method. Chhabra et al. (1995) synthesized anatase TiO_2 nanoparticles (20–50 nm) using $TiCl_4$ and NH_4OH as aqueous phase, Triton X-100 as surfactant, n-hexanol as co-surfactant, cyclohexane as continuous oil phase. The anatase phase had significant performance in the degradation of phenol while the rutile form was inactive for this reaction. It should be noted finally that in microemulsion synthesis as well, the use of other organometallic sources of TiO_2 such as $Ti(OC_4H_9)_4$, and acidity adjustment chemicals such as HCl and HNO_3 is a common practice (Shchukin and Sukhorukov 2004; Wu et al. 1999). The kind of acid used has significant influence on the phase composition of the resulting semiconductor photocatalyst. In a different study, Zurmühl et al. (2011) synthesised nanoscale TiO_2 hollow spheres with 15–20 nm inner diameter, based on gelatine-filled reversed microemulsions.

4.3.6 Combustion Methods

Combustion is one of the suitable routes for the synthesis of nano-TiO_2 particles. In this method of synthesis, a suitable oxidiser is chosen for the stoichiometric combustion of the cation precursor as the combustible. Combustion process has been widely used for the industrial synthesis of TiO_2 since 1958 and currently manufactures multimillion tonnes annually based on the oxidation of purified $TiCl_4$ in pure oxygen flame or plasma at elevated temperatures, typically 1227–1727 °C (West et al. 2007). The overall equation for this plasma-phase reaction is given by Eq. (4.20).

$$TiCl_4 + O_2 \rightarrow TiO_2 + 2Cl_2 \tag{4.20}$$

In small-scale combustion methods extreme conditions are not imposed, so fuel is usually added to the TiO_2 source. The commonest fuel and TiO_2 source are titanyl nitrate $TiO(NO_3)_2$ and glycine H_2NCH_2COOH, heated to about 350 °C and these substances react as follows (Sivalingam et al. 2003; Navageni et al. 2004):

$$9TiO(NO_3)_2 + 10NH_2CH_2CHOOH \rightarrow 9TiO_2 + 25H_2O + 20CO_2 + 14N_2 \tag{4.21}$$

The titanyl nitrate can be synthesised by the reaction of titanyl hydroxide $TiO(OH)_2$ obtained by the hydrolysis of titanium isopropoxide $Ti(i\text{-}OC_3H_7)_4$, with nitric acid (Aarthi and Madras 2008). Urea, thiourea or both can also be used as fuel. For example, ceramic (1-*x*)$TiO_{2-x}SnO_2$ has been prepared by using SnC_2O_4 and $Ti(OC_3H_7)_4$

as metal precursors and urea as fuel (Chinarro et al. 2007). The combustion reaction may be given by Eq. (4.22).

$$SnC_2O_4 + Ti(OC_3H_7)_4 + CO(NH_2)_2 + 20.5O_2 \rightarrow SnO_2\text{-}TiO_2 + 16H_2O + 15CO_2 + N_2 \qquad (4.22)$$

Kitamura et al. (2007) synthesized TiO_2 by the oxidation of Ti metal with $NaClO_4$ (Eq. 4.23). Even though NaCl is formed in the process, it can easily be washed out with water. In the study, the shape and crystal structure of the titanium dioxide product significantly depended on the particle size of the titanium used; the smaller titanium particles (~10 μm) resulted in rutile with an irregular shape, whereas the larger particles (~25 μm) resulted in spheres of anatase. The activity of the rutile-rich sample for methanol and acetic acid decomposition was significantly larger than that of the anatase- rich one in both photocatalytic reactions.

$$Ti + NaClO_4 \rightarrow TiO_2 + NaCl \qquad (4.23)$$

4.3.7 *Miscellaneous Methods*

There are several other methods of synthesis of variety of semiconductor photocatalysts. With the emergence of nano-applications, nanoscopic semiconductor photocatalysts have been constantly produced. Guo et al. (2005) prepared nanoscale TiO_2 powders with narrow size dispersion in supercritical carbon dioxide via non-hydrolytic acylation-deacylation of titanium alkoxide precursors with or without tris-fluorination. With titanium isopropoxide as precursor, such process can be represented by the reaction scheme in Fig. 4.11.

In the non-hydrolytic preparation, titanium isopropoxide, acetic anhydride and titanium (IV) tetrachloride were added to a high-pressure reaction cylinder and liquid CO_2 was then pumped into the vessel. The mixture was stirred for about 30 min at room temperature, and then heated to 110 °C. The pressure was increased to 17 MPa and stirred for 20 h. After the vessel had cooled to room temperature, CO_2 was depressurized, inducing precipitation of a white powder. Small amount of anhydrous hexane was added and the resulting solid powders were collected by centrifugation. The powders were washed with hexane and acetone several times and dried in an oven at 60 °C overnight. Upon calcination at 500 °C for 5 h in air, TiO_2 powder was obtained.

Of the commonest methods for the synthesis of semiconductor photocatalysts, impregnation has become widespread especially for the introduction of dopants and for the fabrication of composite semiconductor photocatalysts. Normally, the photocatalyst material to be doped is suspended while stirring in the impregnated solution for a period long enough to cause impregnation (Surolia et al. 2007). The approach to impregnation may vary depending on the impregnation condition. The resulting impregnated solid material can be obtained by filtration and subsequent drying, or by directly drying the carrier solvent. Due to their convenience, several impregnation

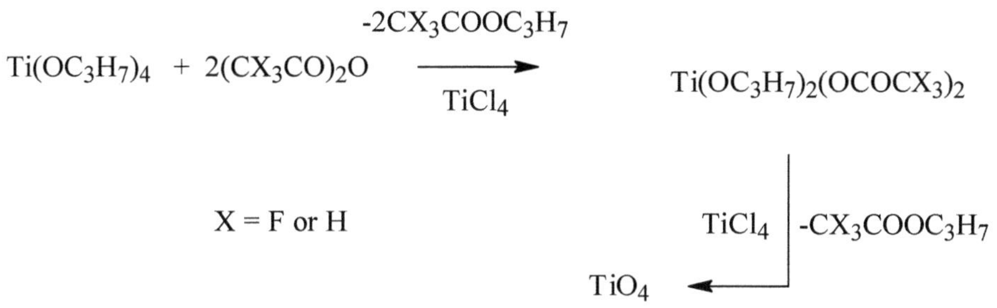

Fig. 4.11 Routes to TiO_2 via non-hydrolytic acylation-deacylation of titanium alkoxide precursors with or without tris-fluorination

methods exist which include wet incipient, supercritical water-based and hydrothermal assisted methods. For instance, by impregnation Byrappa et al. (2006) prepared an inpregnation type zinc oxide-activated carbon composite under mild hydrothermal conditions (150 °C, 20–30 bars).

4.4 Forms of Photocatalyst, Supports and Reactor Configurations

Semiconductor photocatalyst can be contacted with the reactant either in the form of suspension or as coating on a substrate. When used as suspensions, semiconductor photocatalysts provide large surface area which will in turn increase the rate of reaction. One disadvantage of the use of photocatalysts in the form of suspension is that it necessitates the recovery of photocatalyst after treatment (Gumy et al. 2006) which may be inconvenient and unsuitable for cost-effective scale-up. Most of the fundamental and applied studies reported so far utilised semiconductor powders in dispersed form which indicates their convenience for laboratory use. These studies include charge carrier dynamics and mechanistics (Anpo and Kubokawa 1984; Bahnemann et al. 1987; Kamat and Patrick 1992; Colombo et al. 1995). More so, even for the purpose of large scale studies, semiconductor photocatalysts have been used in the dispersed form (Dillert et al. 1999; Giménez et al. 1999).

Certain polymers have sufficient grafting reactivity and lower density to form composites with semiconductor photocatalysts that can float on water. These semiconductor/polymer based composites can therefore provide easier way of photocatalyst recovery. Magalhães et al. (2011) have reported composites of low density polyethylene (LDPE) containing TiO_2 P25 which showed excellent results in the degradation of methylene blue with UV (Hg 245 nm) and solar irradiation under constrained conditions (non-stirring and non-oxygenation). Under the same conditions pure TiO_2 showed very low activities. Such catalysts have potential applicability to the removal of oil spills and treatment of water in open reservoirs. Relatively little is known about the mechanism that led to remarkable improvement in photocatalytic efficiency with floating catalysts. However, one may link such observations to the innovative approach to fix preformed nanocrystalline TiO_2 on low-density polyethylene (LDPE) film, first reported by Zhiyong et al. (2007). The LDPE-TiO_2 film was able to mediate the complete photodiscoloration of Orange II using about seven times less catalyst than a TiO_2 suspension and proceeded with a photonic efficiency of about 0.02. The findings of these workers confirms the high resistance of LDPE-TiO_2 catalyst, full dye discoloration ability under simulated solar light but only to a 30 % total organic carbon (TOC) reduction presumably owing to long-lived intermediates generated in solution and poor mineralisation.

When semiconductor photocatalyst is coated on a substrate the need for free catalyst by physical separation methods is eliminated. However, the major drawback in the use of immobilised photocatalysts is that their active surface area is relatively reduced which in turn diminishes the system efficiency. It has been argued that a reduction of 60–70 % in performance of immobilized TiO_2 may be realised when compared to the unsupported catalyst (Kabra et al. 2004). Various workers have been utilised different substrate surfaces for the semiconductor photocatalyst, which include ceramic tiles (Kemmitt et al. 2004), soda lime glass (Hunoh et al. 2005), aluminium (Chen et al. 2006) and coated glass (Macedo et al. 2007). These substrates are mostly practically inert and do not have any influence on the photocatalytic reaction. For instance, ZnO demonstrated the same efficiency for phenol removal as TiO_2 when both photocatalysts were supported on glass Raschig rings (Yeber et al. 2000). Since coatings are very thin, the actual active surface area of the photoreactor compared to the overall volume is low. In spite of all drawbacks, more coated photocatalysts and immobilisation techniques are still investigated especially with titanium dioxide, a photocatalyst with high surface chemostability. In many of these cases TiO_2 coated on support assumed more efficiency in organic compound removal than uncoated TiO_2 (Kim et al. 2005).

Various materials have been examined for possible use both as catalyst support and to enhance adsorption during photocatalytic processes. These materials include cellulose fibers (Plantard et al. 2011), zeolites (Sampath et al. 1994) and chitosan, layered on glass (Nawi et al. 2012). The use of chitosan with TiO_2 was reported for the first time by Zainal et al. (2009) in the removal of methyl orange. The combination of TiO_2 with chitosan demonstrated photocatalytic effect in consonance with adsorption effect which were attributed to the reactive -NH_2, -OH, and metal oxide in the system. Similarly, zeolites have been used both as supports to concentrate the pollutant (Zainuddin et al. 2010) and to promote TiO_2 dispersion in water (Henderson 2011). Zeolitic materials have potentials for use both in the aqueous and gaseous photocatalytic decontamination. This has usually resulted in significant increase in photocatalytic degradation rate. However, as good adsorbents zeolites themselves may lead parallel removal of the contaminants (Ellis and Korth 1993).

The geometry of reactor is an important factor that can affect photocatalytic efficiency. Inefficiencies in photocatalytic reactors are mainly as a result of light scattering, non-correspondence between the radiation field and fluid residence time, catalyst installation, mass transfer of reactants and reactor kinetics (Puma and Yue 1999; Ray and Beenackers 1998). In order to attain maximum system efficiency and for successful scale up, optimisation of reactor design is imperative. A major challenge in the optimisation of photocatalytic reactor design is the uniform distribution of light flux to a large concentration of the surface area of the photocatalyst. In addition, reactors design or assembly must meet several requirements among which the form the photocatalyst will be used and the phase of the substrate form integral part. Consequently, several reactors have been designed to use slurry or immobilised photocatalyst with artificial light source, or to utilise solar light.

4.4.1 *Slurry Photocatalytic Systems*

Reactors containing photocatalysts in the form of suspension are regarded as slurry or suspension photoreactors. Various slurry photoreactors have been used in photocatalytic decontamination, which include slurry annular reactor, fountain reactor, air-lift loop photoreactor, pulse-baffle tube photochemical reactor, slurry bubble column reactor, rotating tube reactor, concurrent downflow contactor reactor and thin-film reactor (Dijkstra et al. 2003). Customarily, the name of the reactor follows the structural design and/or the method of operation of the reactor. Slurry annular reactor has been among the most extensively used in semiconductor photocatalysis. However, this reactor entails difficulties associated with temperature control, transmission of radiation and the limited amount of catalyst in the reactor (Howe 1998). A representative example of slurry photoreactors is the fountain reactor (Fig. 4.12a). The reactor operates a fountain of the reactivity solution which spreads over an irradiated horizontal plane in a continuous manner. Another example is the slurry bubble photoreactor, which is usually an annular reactor that allows high velocity air bubbles as means of mixing the semiconductor suspension (Pujara et al. 2007). In this reactor as the name implies, air is bubbled in from the bottom. A schematic diagram of slurry bubble reactor is shown in Fig 4.12b.

Slurry photoreactivity systems can be operated in batch or continuous mode. Majority of the photocatalytic studies in print featured cylindrical batch reactor systems which require specified volume of water to be treated in a non-recycled manner. For continuous operation on the other hand, reactors may assemled with a contaminant reservoir in order to permit continuous recirculation of the fluid (Bouchy and Zahraa 2003). This feature is of course more suitable to large scale operations, and in certain reactors will improve system efficiency especially in the event of inappreciable conversion.

4.4.2 *Immobilised Photocatalyst Reactors*

Immobilisation of catalyst allows the use of photocatalytic systems in continuous mode which is advantageous for large scale operation. Several workers have proposed reactors for use with immobilised photocatalysts, which appeared with different names such as coated reactor, rotating disk reactor, glass-mesh reactor, membrane reactor, fluidized-bed reactor, packed bed reactor, fibre optic reactor, hollow-tube lamp reactor and honey-comb monolith reactor (Dijkstra et al. 2003). Reactors may

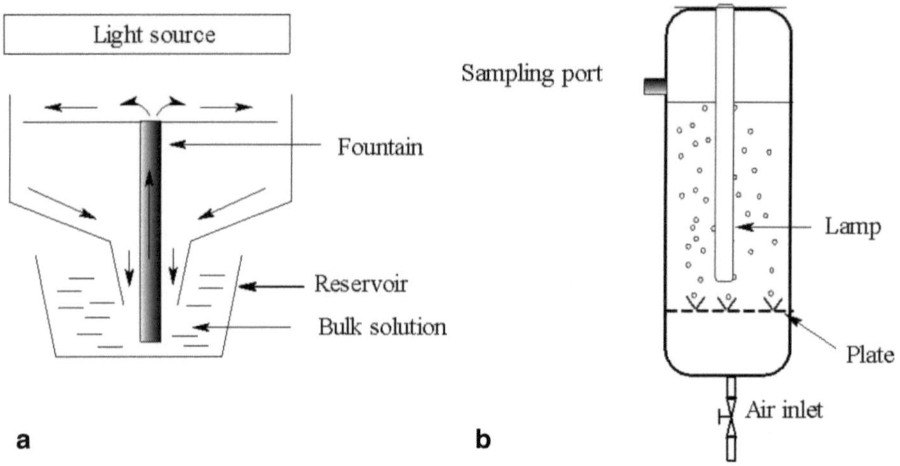

Fig. 4.12 Schematic diagram of suspension reactors (**a**) fountain reactor (**b**) batch bubble column reactor

Fig. 4.13 Schematic diagram of a continuous packed bed photoreactor

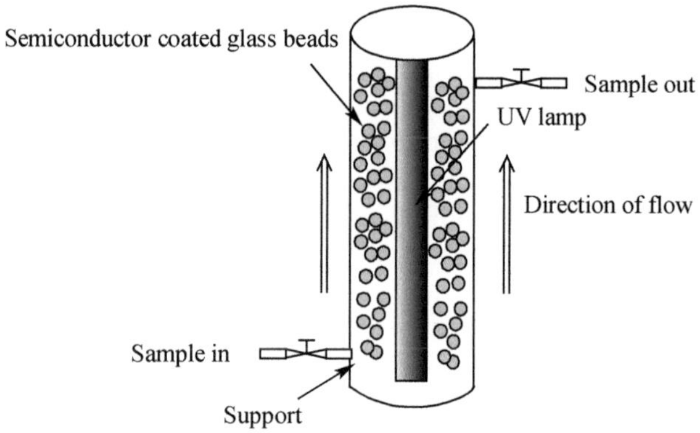

be chosen based on the nature of the pollutant and the medium in which it is present. For example, packed bed photoreactors contain spherical photocatalyst particles coated on inert support such as glass beads, mesh or wool, annularly packed and irradiated by a concentric light source. A schematic diagram of the packed bed photoreactor is shown in Fig. 4.13.

When the packed bed reactor is used for removal of pollutants in solution only limited amount of reactivity solution contacts the irradiated semiconductor photocatalyst. Another disadvantage of this reactor is uneven flow of incident light. Nonetheless, the reactor has been found effective in the photocatalytic oxidation of gaseous contaminants such as ethylene and trichloromethane (Fu et al. 1996; Wang et al. 2002). In photoelectrocatalysis, the packed bed reactor has also found application but in a different form consisting of semiconductor filled electrodes and packed units (An et al. 2004).

4.4.3 Solar Photocatalytic Reactors

Solar-assisted photocatalytic processes currently provide more economically viable and advantageous approach for large scale purpose. Solar photocatalytic processes use only the high-energy short-wavelength photons (<600 nm) to promote photochemical reactions (Blanco-Galvez et al. 2007). Solar photocatalytic reactor can be classified mainly into two: solar concentrating reactors and solar non-concentrating reactors. The performance of both systems has been compared by many workers, and advantages and disadvantages of each one have been established. Solar concentrating reactors consist of solar collectors, which are usually based on glass mirrors designed to track, collect, concentrate and solar light into the reactivity solution. Conversely, non-concentrating are made up of static solar collectors with non-tracking option, thus easier to use or install.

There are five main designs of solar collectors: parabolic trough collector (PC), compound parabolic collector (CPC), tubular collector (TC), inclined plate collector (IC) and V-shaped trough collector (Bandala et al. 2004; Braham and Harris 2009). Tubular collector receives solar light directly without the use of reflectors. The inclined plate collectors are as the name implies, are inclined solar plates within which the reactivity solution flows. This type of reactor has efficiency of solar reception but the control of flow is difficult. The use of this reactor for large scale operations may be impractical as only limited amount of the solution will be allowed into the reactor. The V-shaped trough collector is made of a combination of "V" troughs while the parabolic trough collector and the compound parabolic collector consist of several aluminium-lined parabolic troughs. In the combined parabolic collector transparent absorber tubes between two troughs while in the parabolic trough collector the absorber tubes are aligned within the troughs. A schematic diagram of the structure of a single unit that forms part of combined parabolic collector and pabolic trough collector is shown in Fig. 4.14. A complete module can have eight or more absorber tubes (Parra et al. 2001) spread on an extended parabolic sheet.

Combined parabolic reactor combines all the advantages of both concentrating and non concentrating reactors and none of the disadvantages (Blanco-Galvez et al. 2007). This reactor uses both direct and diffuse light (Herrmann et al. 1998) and requires lesser amount of catalyst. Currently this reactor appears to be more popular. However, the concentration light results in solar thermal photocatalysis which has yet to be extensively investigated.

4.5 Bridging the Photon Transfer Limitations

4.5.1 Dye-Sensitisation

Sensitisation allows extending the wavelength edge of semiconductor particle in the absorption spectrum to visible region. Dye molecules can be facilely excited by visible light to allow transfer of electrons to the conduction band of the TiO_2

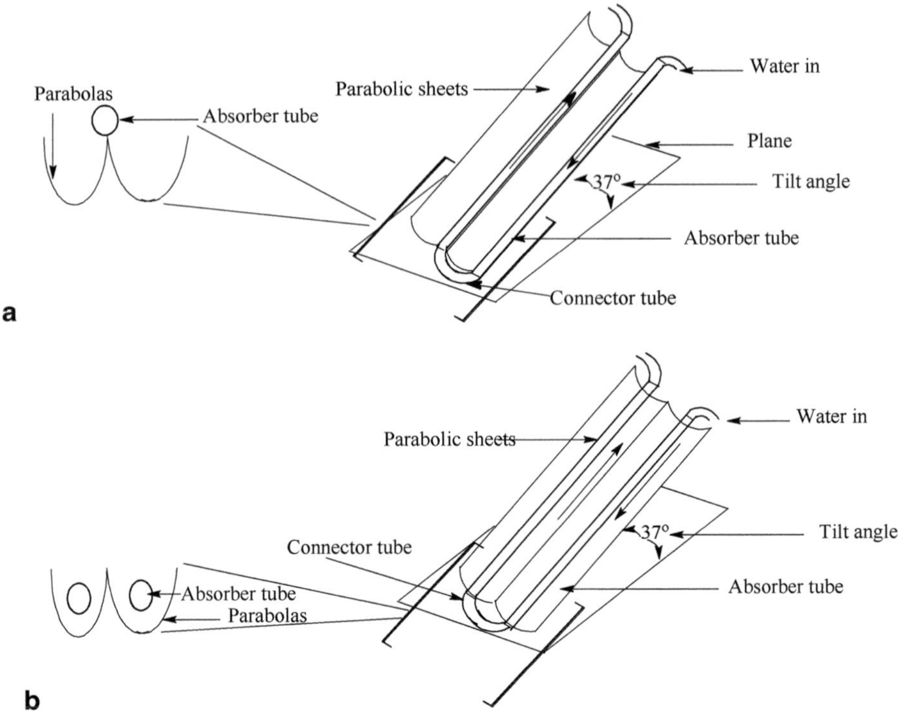

Fig. 4.14 Schematic diagram showing the building block of (**a**) compound parabolic collector, and (**b**) parabolic trough reactor. The position of reactor tube with respect to the parabolic trough in each case has been shown on the *left* hand side

Fig. 4.15 A schematic diagram showing photoinduced dye sensitisation of a semiconductor

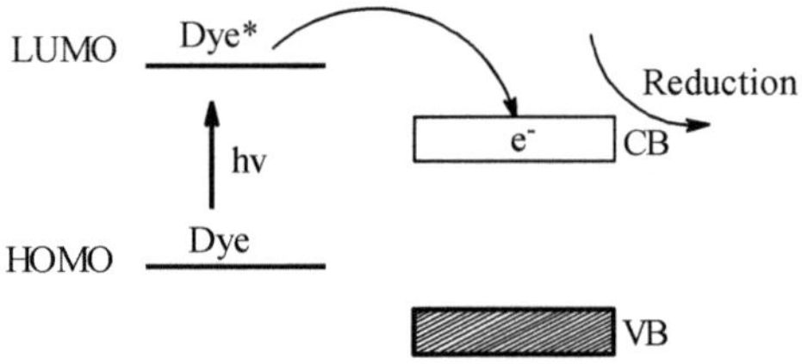

photocatalysts (Yang et al. 2005; Anandan and Yoon 2004). These injected electrons can then be transferred to adsorbed dioxygen to form superoxide anion radicals which lead to the formation of reactive oxygen species which are exploitable in the oxidative abatement of organic molecule. Figure 4.15 shows the electron injection process in dye sensitisation.

As a requirement, the dye must have more negative potential than the conduction band of the semiconductor. Thus, upon irradiation electrons are excited from the highest occupied molecular orbital (HOMO) to a higher energy level in the dye corresponding to the lowest unoccupied molecular orbital (LUMO). These electrons may then be transferred to the conduction band of the semiconductor (SC) for photochemical reaction and the dye is left in a cationic form. The elementary steps involved in the visible light-induced photodegradation of dye by dye-sensitisation mechanism are shown by Eq. (4.22)–(4.26). In the process, the dye gets excited and an electron is injected into the conduction band of the semiconductor.

$$\text{Dye} \xrightarrow{h\nu(\text{vis})} \text{Dye}^* \quad \text{(photoabsorption)} \tag{4.24}$$

$$\text{Dye}^* \rightarrow \text{Dye}^+(\text{e}^-) \quad \text{(dye excitation)} \tag{4.25}$$

$$\text{SC} + \text{Dye}(\text{e}^-) \rightarrow \text{SC}(\text{e}^-{}_{\text{CB}}) + \text{Dye}^+ \quad \text{(electron injection)} \tag{4.26}$$

$$\text{SC}(\text{e}^-{}_{\text{CB}}) + \text{Dye}^+ \rightarrow \text{SC} + \text{Dye} \tag{4.27}$$

$$\text{Dye} + \text{ROS} \rightarrow\rightarrow \text{Photoproducts} \tag{4.28}$$

Dye-sensitised TiO_2 could cause the degradation of pesticides, including also a wide range of chloroaromatics and chloroaliphatics (Chatterjee and Mahata 2001; Chatterjee and Mahata 2004). Indeed, this concept is the basis for the discolouration, degradation and mineralisation of several environmentally important dye compounds (Chen et al. 2002; Dawson and Kamat 2001). The electron transfer pathway can be altered by the addition of metal cations, especially Cu^{2+} (Chen et al. 2002). The disadvantage of dye sensitisation in photocatalytic degradation is that unless the electrons are used in electrical work the dye may also be degraded in the process. In Chap. 5, we have explained how these electrons can be utilised in solar photovoltaic cells.

4.5.2 *Semiconductor/Metal or Non-Metal Nanocomposites*

The incorporation of foreign metal species into semiconductor materials results in a faster transfer of electrons to the conduction band of the semiconductor particles. This increases the charge carrier separation and lessens electron-hole recombination in the semiconductor system. Moreover, the optical absorption of semiconductor photocatalyst can be extended well into the visible region. Basically, the surface of semiconductor particles may be modified to form core/shell heterostructure with improved stability and efficiency. Impurities or dopants can also be introduced into the semiconductor structure in order to narrow the band gap energy of the semiconductor photocatalyst. Dopant can either be metal or non-metal or a combination of both metal and non-metal. Transition metal doping has been a subject of intensive research owing to its potential for wavelength shift. Often in metal doping, zerovalent metal is deposited on the surface of the semiconductor photocatalysts. Therefore, the semiconductor particle serves as a core on which the metal particle or shell may be capped. Figure 4.16 shows a core/shell model representing a semiconductor/metal heterostructure.

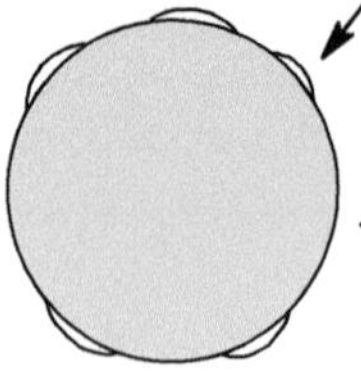

Fig. 4.16 Schematic model showing semiconductor/metal heterostructure

Catalysis with coinage and platinum metals is currently visited with renewed interest. Dawson and Kamat (2001) have demonstrated the synthesis of nano-TiO_2/Au nanocomposite by reduction of $[AuCl_4]^-$ over TiO_2 to form Au/TiO_2. The efficacy of nanogold-modified TiO_2 surface for water splitting, hydrogen production from ethanol (Yuzawa et al. 2012) and methanol conversion to formaldehyde (Ismail et al. 2009) has been established. Several Ag/ZnO heterostructure nanocatalysts with excellent photocatalytic performance have also been successfully prepared by different methods (Zheng et al. 2007; Moonsiri et al. 2004). Enhanced photocatalytic activities of the synthesized Ag/ZnO nanocrystallites were clearly observed during the photocatalysis of methylene blue and phenol (Wang et al. 2004). Platinum-doped semiconductor materials are also known to significantly absorb visible light and lead to extended band gap energies. Significant drop in band gap energy from 3.11 eV for pure TiO_2 to 2.95 eV for Pt (1.5wt %)/TiO_2 (Chen et al. 2007). Similarly, ZnO powders loaded with platinum metal Pd (or Ag) has demonstrated enhancement in the photocatalytic degradation of gas-phase n-heptane (or rhodamine B) (Liqiang et al. 2004; Zhang and Mu 2007).

The efficiency of Pt-loaded TiO_2 may be explained based on work function (Φ), defined as the energy required to remove an electron from Fermi energy level of the surface to the vacuum level. For TiO_2, work function is given by Eq. (4.29).

$$\Phi_{TiO_2} = \Phi_i + \Phi_\delta + \chi \tag{4.29}$$

Where Φ_{TiO_2} is the work function of TiO_2, Φ_i is the internal work function, Φ_δ is the work function related to band bending and χ is the external work function. Upon consideration of many reports the following conclusions about Φ_{TiO_2} and Φ_{Pt} were drawn (Nowotny et al. 2008):

$$2.9\text{eV}\Phi_{TiO_2} < 3.2\text{eV}$$

$$5.12\text{eV}\Phi_{Pt} < 5.93\text{eV}$$

It follows that that $\Phi_{Pt} > \Phi_{TiO_2}$, which shows the significance of Pt loading in the preparation of TiO_2.

Principally, the cationic doping of TiO_2 particles is based on the induction valence law which implies that trivalent cations of suitable sizes can occupy Ti^{4+} lattice and behave as electron acceptor centres (Herrmann et al. 1984). Some rare earth metals from both cerium group (which includes lanthanum La, cerium Ce, praseodymium

Pr, neodymium Nd, promethium Pm, samarium Sm, europium Eu) and yttrium group (which includes yttrium Y, godolinium Gd, terbium Tb, dysprosium Dy, holmium Ho, erbium Er, thulium Tm, Ytterbium Yb, lutetium Lu) have been identified as excellent dopants for use in photocatalyst synthesis. Lanthanum has received special interest due to its superior applicability as dopant (Uzunova-Bujnova et al. 2008). Recently, Anandan et al. (2007) have shown the superior photoactivity of La-doped ZnO in the removal of 2,4,6-trichlorophenol, a model pollutant. Similarly, the presence of samarium as dopant, at optimum amount of 0.5 mol %, effectively inhibited the recombination of electron and hole and improved the activity of TiO_2 for methylene blue degradation (Xiao et al. 2008). For the purpose of comparison, the photoelectrochemical and photocatalytic properties of lanthanide metal ion-doped TiO_2 nanoparticles were investigated by Wang et al. (2000) and the results showed that the photoresponse of Eu^{3+} -, La^{3+} -, Nd^{3+} -and Pr^{3+} -doped TiO_2 electrodes were much larger, and that of Sm^{3+} -doped TiO_2 electrode was a little larger than that of undoped TiO_2 electrode, indicating that the photogenerated carriers were separated more efficiently in Eu^{3+} -, La^{3+} -, Nd^{3+} - and Pr^{3+} -doped TiO_2 nanoparticles than in undoped TiO_2 nanoparticles. About 44 % of the solar spectrum is in the near-infrared (NIR) region but less work has been reported utilising this spectral reagion. Qin et al. (2010) therefore have for the first time demonstrated the potentials of NIR-photocatalysis of lanthanoids using YF_3: Yb^{3+}, Tm^{3+}/TiO_2 core/shell nanoparticles and methylene blue as model pollutant.

A number of studies reported doping with anions such as I, C, N and S. When used as dopant, nitrogen can remarkably extend the spectral absorption of semiconductor photocatalysts. Titanium dioxide (110) single crystals, doped with nitrogen via an ammoniacal treatment at 597 °C, have been found to exhibit photoactivity at photon energies down to 2.4 eV, which is 0.6 eV below the band-gap energy for rutile TiO_2 (Diwald et al. 2004). A co-doping effect between nitrogen and hydrogen is postulated to be responsible for the enhanced photoactivity of nitrogen-doped TiO_2 materials in the range of visible light spectrum. Films and powders of $TiO_{2-x}N_x$ have shown remarkable improvement in optical absorption and photocatalytic activity under visible light (Asahi et al. 2001). Owing to these attributes, N-doped semiconductor photocatalysts have attracted intensive research (Jiang et al. 2008). Carbon-doped semiconductor photocatalysts are also highly effective in the decontamination of persistent organic compounds (Lim et al. 2008). It has been observed that the TiO_2/C composites with 5.8 mass % or higher carbon content have both poor adsorption capability and low photocatalytic activity (Li et al. 2008). However, TiO_2/C containing 1.2 mass % carbon exhibits both excellent adsorption capability and photocatalytic activity against methyl orange. These studies indicate the necessity for optimisation of carbon load in TiO_2/C composites.

Unique synergy may be driven by binary doping which can take the form of metal-nonmetal, metal-metal or non-metals binary doping. In fact it became imperative for the following reasons. Mrowetz et al. (2004) have seen the failure of N-doped TiO_2 to catalyse the oxidation of simple $HCOO^-$ into $CO_2^{\bullet}$ -, or of NH_3-OH^+ into NO_3^-, under visible illumination (λ>400 nm) either by direct means or via intermediate species produced in the oxidation of water or the catalyst. Secondly, the efficacy of mono-metal doping is still questioned. Moreover, superior catalytic activity was observed in the La-N codoped TiO_2 under visible light ($350<\lambda<450$ nm) than realised with either N- or La-doped TiO_2 (Wei et al. 2004).

Newer approaches for the enhancement of photocatalytic properties include the establishment of p-n junction heterostructure in the semiconductor photocatalyst. Nano p-n junction of TiO_2 nanobelt heterostructures can be produced by assembling *p*-type semiconductor (such as NiO) particles on *n*-type semiconductor photocatalyst (such as TiO_2) in order to enhance activity for solar photocatalysis (Lin et al. 2007).

4.5.3 Downsizing

The size of semiconductor particle is crucial to its photocatalytic activity. Bulk materials usually have uniform properties irrespective of particle size. However, as the diameter of a semiconductor particle diminishes down to ultrafine size regimes there is modification in physical properties due to what is known generally referred to as quantum size effect. It would be noted that for almost all the subsecond laser and photoluminescence experiments on charge carrier activities dispersions of nanocatalyst particles were utilised due to their smaller dimension compared to the wavelength of irradiation, which provides for better absorption during study (Moser et al. 1985; Tachibana et al. 1996). Cavaleri et al. (1995) studied the influence of size between 3.2 and 6.2 nm on the electronic relaxation processes of quantum sized ZnO using femtosecond spectroscopy. The study showed that the larger the cluster size the faster the electron-hole trapping and recombination.

In addition to the foregoing charge carrier events and luminescence, downsizing affects the band gap of semiconductor, surface area and structure. Zinc oxide is usually an *n*-type semiconductor with direct band gap between 3.2 and 3.4 at room temperature depending on preparation (Bhat and Deepak 2005; Marotti et al. 2006). Generally, a shift to higher energies (i.e. blue shift) is observed as semiconductor is reduced to smaller nanosizes due to confinement effect (Anpo et al. 1987). In a study of the performance of ultrafine ZnO of 12, 18 and 25 nm in phenol degradation, increase in the photocatalytic activities of the particles was observed with reduction in size (Jing et al. 2001). Similarly, Maira et al. (2001) investigated the influence of the TiO_2 particle size in the gas phase photocatalytic oxidation of toluene using nanostructured TiO_2 catalysts of controlled particle size: 6, 11, 16 and 20 nm. They observed higher conversion and complete mineralization with the smallest particle size, 6 nm. The effect of downsizing is observed regardless of whether the photocatalysis is in liquid-solid regime or gas-solid regime. Consequently, quantum dots are now used widely to improve photocatalytic energy production, photocatalytic degradation and self-cleaning materials.

4.6 Conclusion

Several factors may dictate the efficiency of semiconductor photocatalyst in photocatalytic processes. Majority of these factors however are oriented around ensuring, larger light absorption by the photocatalyst, reduced recombination of charge carriers and surface stability. It is therefore desirable to fabricate semiconductor photocatalysts that can overcome as many shortcomings as possible. Currently, there is no shortage of methods of synthesis of semiconductor photocatalysts of different size scale and quality. Therefore, not only efficient catalysts are required but simultaneous achievement of high performance when these preparations are used in scale-up projects.

The design of an efficient photocatalytic system for successful scale-up may be a difficult task involving the design or suitable reactors, lamp technology and optimisation of several variables. At present, developments in solar photocatalytic reactor design are under way towards ensuring cost-effective and efficient photocatalytic systems. The utilisation of experimental design, engineering and simulative methods may bring improvement in this line. Finally, increasing sophistication in analytical instruments and reactor design is still needed to further revolutionise both the synthesis of photocatalysts and their characterisation for significant overall efficiency of the available semiconductor photocatalytic processes.

References

Aarthi T, Madras G (2008) Photocatalytic reduction of metals in presence of combustion synthesized nano-TiO_2. Catal Commun 9:630–634

Abdollahi Y, Abdullah AH, Gaya UI et al (2011), Enhanced photodegradation of o-cresol in aqueous Mn(1 %)-doped ZnO suspensions. Environ Technol 33:1183–1189

An TC, Zhang WB, Li GY et al (2004) Increase the degradation efficiency of organic pollutants with a radical scavenger (Cl^-) in a novel photoelectrocatalytic reactor. Chin Chem Lett 15:455–458

Anandan S, Yoon M (2004) Photoinduced electron transfer studies of Nile red in the presence of TiO_2 colloidal nanoparticles. Spectrochim Acta Part A 60:885–888

Anandan S, Vinu A, Mori T et al (2007) Photocatalytic degradation of 2,4,6-trichlorophenol using lanthanum doped ZnO in aqueous suspension. Catal Commun 8:1377–1382

Anpo M, Kubokawa Y (1984) Photoluminescence of zinc oxide powder as a probe of electron-hole surface processes. J Phys Chem 88:5556–5560

Anpo M, Shima T, Kodama S (1987) Photocatalytic hydrogenation of CH_3CCH with H_2O on small-particle TiO_2: size quantization effects and reaction intermediates. J Phys Chem 97:4305–4310

Ardizzone S, Pirola C, Bianchi CL (2007) Tailored anatase/brookite nanocrystalline TiO_2. The optimal particle features for liquid and gas-phase photocatalytic reactions. J Phys Chem C 111:13222–13231

Arlt T, Bermejo M, Blanco MA et al (2000) High-pressure polymorphs of anatase TiO_2. Phys Rev B 51(61):14415–14419

Asahi R, Morikawa T, Ohwaki T (2001) Visible-light photocatalysis in nitrogen-doped titanium oxides. Science 293:269–271

Augugliaro V, Loddo V, Palmisano G et al (2010) Clean by light irradiation practical applications of supported TiO_2. Royal Society of Chemistry, United Kingdom

Ayllón JA, Figueras A, Garelik S et al (1999) Preparation of TiO_2 powder using titanium tetraisopropoxide decomposition in a plasma enhanced chemical vapor deposition (PECVD) reactor. J Mater Sci Lett 18:1319–1321

Bacsa RR, Kiwi J (1998) Effect of rutile phase on the photocatalytic properties of nanocrystalline titania during the degradation of *p*-coumaric acid. Appl Catal B 16:19–29

Bahnemann DW, Kormann C, Hoffmann MR (1987) Preparation and characterization of quantum size zinc oxide: a detailed spectroscopic study. J Phys Chem 91:3789–3798

Baiju KV, Shukla S, Biju S et al (2009) Morphology-dependent dye-removal mechanism as observed for anatase-titania photocatalyst. Catal Lett 131:663–671

Bandala ER, Arancibia-Bulnes CA, Orozco SL et al (2004) Solar photoreactors comparison based on oxalic acid photocatalytic degradation. Sol Energy 77:503–512

Bandara J, Tennakone K, Jayatilaka PPB (2002) Composite Tin and Zinc oxide nanocrystalline particles for enhanced charge separation in sensitized degradation of dyes. Chemosphere 49:439–445

Bessergenev VG, Pereira RJF, Mateus MC et al (2006) Study of physical and photocatalytic properties of titanium dioxide thin films prepared from complex precursors by chemical vapour deposition. Thin Solid Films 503:29–39

Bhat SV, Deepak FL (2005) Tuning the bandgap of ZnO by substitution with Mn^{2+}, Co^{2+} and Ni^{2+}. Solid State Commun 135:345–347

Blanco-Galvez J, Fernández-Ibáez P, Malato-Rodríguez S (2007) Solar photocatalytic detoxification and disinfection of water: recent overview. J Sol Energy Eng 129:4–15

Bouchy M, Zahraa O (2003) Photocatalytic reactors. Int J Photoenergy 5:191–197

Boxer RJ (1997) Essentials of organic chemistry. WCB/McGraw-Hill Companies, USA

Braham RJ, Harris AT (2009) Review of major design and scale-up considerations for solar photocatalytic reactors. Ind Eng Chem Res 48:8890–8905

Brinker CJ, Scerrer GW (1990) Sol-gel science: the physics and chemistry of sol-gel processing. Academic Press Inc., San Diego

Brunauer S, Emmett PH, Teller E (1938) Adsorption of gases in multimolecular layers. J Am Chem Soc 60:309–319

Byrappa K, Subramani AK, Ananda S et al (2006) Impregnation of ZnO onto activated carbon under hydrothermal conditions and its photocatalytic properties. J Mater Sci 41:1355–1362

Calza P, Pelizzetti E (2001) Photocatalytic transformation of organic compounds in the presence of inorganic ions. Pure Appl Chem 73:1839–1848

Carp O, Huisman CL, Reller A (2004) Photoinduced reactivity of titanium dioxide, progress in solid state chemistry. Prog Solid State Chem 32:33–177

Cavaleri JJ, Skinner DE, Colombo DP Jr (1995) Femtosecond study of the size-dependent charge carrier dynamics in ZnO nanocluster solutions. J Chem Phys 103:5378

Chatterjee D, Mahata A (2001) Demineralization of organic pollutants on the dye modified TiO_2 semiconductor particulate system using visible light. Appl Catal B 33:119–125

Chatterjee D, Mahata A (2004) Evidence of superoxide radical formation in the photodegradation of pesticide on the dye modified TiO_2 surface using visible light. J Photochem Photobiol A 165:19–23

Chen C, Li X, Ma W (2002) Effect of transition metal ions on the TiO_2-assisted photodegradation of dyes under visible irradiation: a probe for the interfacial electron transfer process and reaction mechanism. J Phys Chem B 106:318–324

Chen S-Z, Zhang P-Y, Zhu W-P (2006) Deactivation of TiO_2 photocatalytic films loaded on aluminium: XPS and AFM analyses. Appl Surf Sci 252:7532–7538

Chen H-W, Ku Y, Kuo YL (2007) Effect of Pt/TiO_2 characteristics on temporal behavior of o-cresol decomposition by visible light-induced photocatalysis. Water Res 41:2069–2078

Chhabra V, Pillai V, Mishra BK et al (1995) Synthesis, characterization, and properties of microemulsion-mediated nanophase TiO_2 particles. Langmuir 11:3307–3311

Chinarro E, Moreno B, Jurado JR (2007) Combustion synthesis and EIS characterization of TiO_2–SnO_2 system. J Eur Ceram Soc 27:3601–3604

Clayden J, Warren S, Greeves N (2001) Organic chemistry. Oxford University Press, New York

Colombo DP Jr, Skinner DE, Cavaleri JJ et al (1995) Femtosecond spectroscopy of quantum- and bulk-sized semiconductor particles. In: Abstracts of the 210th ACS National Meeting. United States of America, 20–24 August, Chicago

Comparelli R, Fanizza E, Curri ML et al (2005a) UV-induced photocatalytic degradation of azo dyes by organic-capped ZnO nanocrystals immobilized onto substrates. Appl Catal B 60:1–11

Comparelli R, Fanizza E, Curri ML et al (2005b) Photocatalytic degradation of azo dyes by organic-capped anatase TiO_2 nanocrystals immobilized onto substrates. Appl Catal B 55:81–91

Cun W, Jincai Z, Xinming W et al (2002) Preparation, characterization and photocatalytic activity of nano-sized ZnO/SnO_2 coupled photocatalysts. Appl Catal B 39:269–279

Dawson A, Kamat PV (2001) Semiconductor-metal nanocomposites. Photoinduced fusion and photocatalysis of gold-capped TiO_2 (TiO_2/Gold) nanoparticles. J Phys Chem B 105:960–966

Degen A, Kosec M (2003) Influence of pH and ionic impurities on the adsorption of poly(acrylic) dispersant onto a zinc oxide surface. J Am Ceram Soc 86:2001–2010

Diaz J, Rodríguez J, Ponce S et al (2007) Solar photocatalytic decontamination of phenol using pyrolytic TiO_2 films deposited inside glass tubing. J Sol Energy Eng 129:94–99

Dijkstra MFJ, Koerts ECB, Beenackers AACM et al (2003) Performance of immobilized photocatalytic reactors in continuous mode. AIChE J 49:734–744

Dillert R, Cassano AE, Goslich R et al (1999) Large scale studies in solar catalytic wastewater treatment. Catal Today 54:267–282

Ding Z, Hu X, Lu GQ et al (2000) Greenfield, novel silica gel supported TiO_2 photocatalyst synthesized by CVD method. Langmuir 16:6216–6222

Ding Y, Zhang G, Wu H et al (2001) Nanoscale magnesium hydroxide and magnesium oxide powders: control over size, shape, and structure via hydrothermal synthesis. Chem Mater 13:435–440

Diwald O, Thompson TL, Zubkov T et al (2004) Photochemical activity of nitrogen-doped rutile TiO_2(110) in visible light. J Phys Chem B 108:6004–6008

Dubrovinskaia NA, Dubrovinsky LS, Ahuja R et al (2001) Experimental and theoretical identification of a new high-pressure TiO_2 polymorph. Phys Rev Lett 87:1–4

Ellis J, Korth W (1993) Removal of geosmin and methylisoborneol from drinking water by adsorption on ultrastable zeolite-Y. Water Res 27:535–539

Emeline AV, Smirnova LG, Kuzmin GN et al (2002) Spectral dependence of quantum yields in gas–solid heterogeneous photosystems Influence of anatase/rutile content on the photostimulated adsorption of dioxygen and dihydrogen on titania. J Photochem Photobiol A 148:97–102

Fierro JLG (2006) Metal oxides. Chemistry and applications. Taylor and Francis, USA

Fu X, Clark LA, Zeltner WA et al (1996) Effects of reaction temperature and water vapour content on the heterogeneous photocatalytic oxidation of ethylene. J Photochem Photobiol C 97:181–186

Fujishima A, Rao TN (1997) Recent advances in heterogeneous photocatalysis. Proc Indian Acad Sci (Chem Sci) 109:471–486

Ganguli AK, Ganguly A, Vaidya S (2010) Microemulsion-based synthesis of nano-crystalline materials. Chem Soc Rev 39:474–485

Gaya UI, Abdullah AH, Zainal Z et al (2009) Photocatalytic treatment of 4-chlorophenol in aqueous ZnO suspensions: intermediates, influence of dosage and inorganic anions. J Hazard Mater 168:57–63

Gaya UI, Abdullah AH, Hussein MZ et al (2010) Photocatalytic removal of 2,4,6-trichlorophenol from water exploiting commercial ZnO powder. Desalination 263:176–182

Giménez J, Curcó D, Queral MA (1999) Photocatalytic treatment of phenol and 2,4-dichlorophenol in a solar plant in the way to scaling-up. Catal Today 54:229–243

Gouma PI, Mills MJ (2001) Anatase-to-rutile transformation in titania powders. J Am Ceram Soc 84:619–22

Gumy D, Rincon AG, Hajdu R et al (2006) Solar photocatalysis for detoxification and disinfection of water: different types of suspended and fixed TiO_2 catalysts study. Sol Energy 80:1376–1381

Guo G, Whitesell JK, Fox MA (2005) Synthesis of TiO_2 photocatalysts in supercritical CO_2 via a non-hydrolytic route. J Phys Chem B 109:18781–18785
Helmersson U, Lattemann M, Bohlmark J et al (2006) Ionized physical vapor deposition (IPVD): a review of technology and applications. Thin Solid Films 513:1–24
Hench LL, West JK (1990) The sol-gel process. Chem Rev 90:33–72
Henderson MA (2011) A surface science perspective on TiO_2 photocatalysis. Surf Sci Rep 66: 185–297
Herrmann J-M, Disdier J, Pichat P (1984) Effect of chromium doping on the electrical and catalytic properties of powder titania under UV and visible illumination. Chem Phys Lett 108:618–622
Herrmann JM, Disdier J, Pichat P et al (1998) TiO_2-based solar photocatalytic detoxification of water containing organic pollutants. Case studies of 2,4-dichlorophenoxyaceticacid (2,4-D) and of benzofuran. Appl Catal B 17:15–23
Hoffmann MR, Martin ST, Choi W et al (1995) Environmental applications of semiconductor photocatalysis. Chem Rev 95:69–96
Howe RF (1998) Recent developments in photocatalysis. Dev Chem Eng Mineral Process 6:55–84
Hsiao K-C, Liao S-C, Chen Y-J et al (2007) Synthesis, characterization and photocatalytic property of nanostructured Al-doped ZnO powders prepared by spray pyrolysis. Mater Sci Eng 447:71–76
Hsu C-C, Wu NL (2005) Synthesis and photocatalytic activity of ZnO/ZnO_2 composite. J Photochem Photobiol A 172:269–274
Huang SS, Chen JS (2002) Comparison of the characteristics of TiO_2 thin films prepared by low-pressure and plasma-enhanced chemical vapour deposition. J Mater Sci Mater Electron 13:77–81
Hunoh S, Kim JS, Chung JS et al (2005) Crystallization and photoactivity of TiO_2 films formed on soda lime glass by a sol-gel dip-coating process. Chem Eng Comm 192:327–335
Hurum DC, Agrios AG, Gray KA et al (2003) Explaining the enhanced photocatalytic activity of Degussa P25 mixed-phase TiO_2 using EPR. J Phys Chem B 107:4545–4549
Ismail AA, Bahnemann DW, Bannat I (2009) Gold nanoparticles on mesoporous interparticle networks of titanium dioxide nanocrystals for enhanced photonic efficiencies. J Phys Chem C 113:7429–7435
Jang JS, Yu C-J, Choi SH et al (2008) Topotactic synthesis of mesoporous ZnS and ZnO nanoplates and their photocatalytic activity. J Catal 254:144–155
Jiang X, Wang T (2007) Influence of preparation method on morphology and photocatalysis activity of nanostructured TiO_2. Environ Sci Technol 41:4441–4446
Jiang Z, Yang F, Luo N et al (2008) Solvothermal synthesis of N-doped TiO_2 nanotubes for visible-light-responsive photocatalysis. Chem Commun 2008:6372–6374
Jing L, Xu Z, Sun X (2001) The surface properties and photocatalytic activities of ZnO ultrafine particles. Appl Surf Sci 180:308–314
Kabra K, Chaudhary R, Sawhney RL (2004) Treatment of hazardous organic and inorganic compounds through aqueous-phase photocatalysis. Ind Eng Chem Res 43:7683–7696
Kajbafvala A, Ghorbani H, Paravar A et al (2012) Effects of morphology on photocatalytic performance of zinc oxide nanostructures synthesized by rapid microwave irradiation methods. Superlattices Microstruct 51:512–522
Kallay N, Babić D, Matijević E (1986) Adsorption at solid/solution interfaces II. Surface charge and potential of spherical colloidal titania. Colloids Surf 19:375–386
Kamat PV, Patrick B (1992) Photophysics and photochemistry of quantized ZnO colloids. J Phys Chem 96:6829–6834
Kandiel TA, Feldhoff A, Robben L et al (2010) Tailored titanium dioxide nanomaterials: anatase nanoparticles and brookite nanorods as highly active photocatalysts. Chem Mater 22:2050–2060
Kao L-H, Hsu T-C, Lu H-Y (2007) Sol–gel synthesis and morphological control of nanocrystalline TiO_2 via urea treatment. J Colloid Interface Sci 316:160–167
Kemmitt T, Al-Salim NI, Waterland M et al (2004) Photocatalytic titania coatings. Curr Appl Phys 4:189–192
Kim H, Lee S, Han Y (2005) Preparation of dip-coated TiO_2 photocatalyst on ceramic foam pellets. J Mater Sci 40:5295–5298

Kim C, Doh SJ, Lee SG et al (2007) Visible-light absorptivity of a zincoxysulfide (ZnO_xS_{1--x}) composite semiconductor and its photocatalytic activities for degradation of organic pollutants under visible-light irradiation. Appl Catal A 330:127–133

Kitamura Y, Okinaka N, Shibayama T, Akiyama T et al (2007) Combustion synthesis of TiO_2 nanoparticles as photocatalyst. Powder Technol 176:93–98

Klug HP, Alexander LE (1974) X-Ray diffraction procedures for polycrystalline and amorphous materials. Wiley, New York

Kominami H, Murakami S-Y, Kato J-I (2002) Correlation between some physical properties of titanium dioxide particles and their photocatalytic activity for some probe reactions in aqueous systems. J Phys Chem B 106:10501–10507

Kominami H, Kato J-I, Murakami S-Y et al (2003) Solvothermal syntheses of semiconductor photocatalysts of ultra-high activities. Catal Today 84:181–189

Kuo C-L, Kuo T-J, Huang MH (2005) Hydrothermal Synthesis of ZnO microspheres and hexagonal microrods with sheetlike and platelike nanostructures. J Phys Chem B 109:20115–20121

Lathasree S, Rao AN, SivaSankar B (2004) Heterogeneous photocatalytic mineralisation of phenols in aqueous solutions. J Mol Catal A Chem 223:101–105

Lei S, Duan W (2008) Highly active mixed-phase TiO_2 photocatalysts fabricated at low temperature and the correlation between phase composition and photocatalytic activity. J Environ Sci 20:1263–1267 (Beijing, China)

Li D, Haneda H (2003) Morphologies of zinc oxide particles and their effects on photocatalysis. Chemosphere 51:129–137

Li G, Gray KA (2007) Preparation of mixed-phase titanium dioxide nanocomposites via solvothermal processing. Chem Mater 19:1143–1146

Li J-G, Ishigaki T, Sun X (2007) Anatase, brookite, and rutile nanocrystals via redox reactions under mild hydrothermal conditions: phase-selective synthesis and physicochemical properties. J Phys Chem C 111:4969–4976

Li Z, Xiong Y, Xie Y (2003) Selected-control synthesis of zno nanowires and nanorods via a PEG-assisted route. Inorg Chem 42:8105–8109

Li M, Zhou S, Zhang Y (2008) One-step solvothermal preparation of TiO_2/C composites and their visible-light photocatalytic activities. Appl Surf Sci 254:3762–3766

Liao S, Donggen H, Yu D et al (2004) Preparation and characterization of ZnO/TiO_2, SO_4^{2-}/ZnO/TiO_2 photocatalyst and their photocatalysis. J Photochem Photobiol A 168:7–13

Lim M, Zhou Y, Wood B (2008) Fluorine and carbon codoped macroporous titania microspheres: highly effective photocatalyst for the destruction of airborne styrene under visible light. J Phys Chem C 112:19655–19661

Lin J, Shen J, Wang R (2011) Nano-p–n junctions on surface-coarsened TiO_2 nanobelts with enhanced photocatalytic activity. J Mater Chem 21:5106–5113

Liqiang J, Baiqi W, Baifu X et al (2004) Investigations on the surface modification of ZnO nanoparticle photocatalyst by depositing Pd. J Solid State Chem 177:4221–4227

Liu B, Zeng HC (2003) Hydrothermal synthesis of ZnO nanorods in the diameter regime of 50 nm. J Am Chem Soc 125:4430–4431

Lokhande CD, Lee EH, Jung KD et al (2004) Room temperature chemical deposition of amorphous TiO_2 thin films from Ti(III) chloride solution. J Mater Sci 39:2915–2918

Macedo LC, Zaia DAM, Moore GJ et al (2007) Degradation of leather dye on TiO_2: a study of applied experimental parameters on photoelectrocatalysis. J Photochem Photobiol A 185:86–93

Maeda Y, Fujisjima A, Honda K (1982) The investigation of energy efficiency of the ZnO semiconductor photoelectrode by photothermal measurement. Bull Chem Soc Jpn 55:3372–3376

Magalhães F, Moura FCC, Lago RM (2011) TiO_2/LDPE composites: a new floating photocatalyst for solar degradation of organic contaminants. Desalination 276:266–271

Maira AJ, Yeung KL, Soria J (2001) Gas-phase photo-oxidation of toluene using nanometer-size TiO_2 catalysts. Appl Catal B 29:327–336

Marotti RE, Giorgi P, Machado G (2006) Crystallite size dependence of band gap energy for electrodeposited ZnO grown at different temperatures. Sol Energy Mater Sol Cells 90:2356–2361

Mills A, Elliott N, Parkin IP et al (2002) TiO_2 CVD films for semiconductor photocatalysis. J Photochem Photobiol A 151:171–179
Mo S-D, Ching WY (1995) Electronic and optical properties of three phases of titanium dioxide: rutile, anatase, and brookite. Phys Rev B 51:13023–13032
Moonsiri M, Rangsunvigit P, Chavadej S et al (2004) Effects of Pt and Ag on the photocatalytic degradation of 4-chlorophenol and its by-products. Chem Eng J 97:241–248 (Amsterdam, Neth)
Morkoç H, Özgür Ü (2009) Zinc Oxide: fundamentals, materials and device technology. Wiley-VCH Verlag GmbH & Co. KGaA, Weinheim
Moser J, Grätzel M, Sharma DK et al (1985) Picosecond time-resolved studies of photosensitised electron injection in colloidal semiconductors. Helv Chim Acta 68:1686–1690
Moser J, Gratzel M, Gallay R (1987) Inhibition of electron-hole recombination in substitutionally doped colloidal semiconductor crvstallites. Helv Chim Acta 70:1596–1604
Mrowetz M, Balcerski W, Colussi AJ (2004) Oxidative power of nitrogen-doped TiO_2 photocatalysts under visible illumination. J Phys Chem B 108:17269–17273
Navageni K, Sivalingam G, Hegde MS et al (2004) Photocatalytic degradation of organic compounds over combustion-synthesized nano-TiO_2. Environ Sci Technol 38:1600–1604
Nawi MA, Sabar S, Sheilatina (2012) Photocatalytic decolourisation of reactive red 4 dye by an immobilised TiO_2/chitosan layer by layer system. J Colloid Interface Sci 372:80–87
Neppolian B, Jung H, Choi H (2007) Photocatalytic Degradation of 4-chlorophenol using TiO_2 and Pt-TiO_2 Nanoparticles prepared by sol-gel method. J Adv Oxid Technol 10(2):001–006
Nowotny MK, Sheppard LR, Bak T et al (2008) Defect chemistry of titanium dioxide. application of defect engineering in processing of TiO_2-based photocatalysts. J Phys Chem C 112:5275–5300
Ohno T, Sarukawa K, Tokieda K et al (2001a) Morphology of a TiO_2 photocatalyst (Degussa, P-25) consisting of anatase and rutile crystalline phases. J Catal 203:82–86
Ohno T, Sarukawa K, Matsumura M (2001b) Photocatalytic activities of pure rutile particles isolated from TiO_2 powder by dissolving the anatase component in HF solution. J Phys Chem B 105:2417–2420
Paola AD, Cufalo G, Addamo M, Bellardita M, Campostrini R, Ischia M, Ceccato R, Palmisano L (2008) Photocatalytic activity of nanocrystalline TiO_2 (brookite, rutile and brookite-based) powders prepared by thermohydrolysis of $TiCl_4$ in aqueous chloride solutions. Colloids Surf A 317:366–376
Parida KM, Parija S (2006) Photocatalytic degradation of phenol under solar radiation using microwave irradiated zinc oxide. Sol Energy 80:1048–1054
Parra S, Malato S, Blanco J et al (2001) Concentrating versus non-concentrating reactors for solar photocatalytic degradation of p-nitrotoluene-o-sulfonic acid. Water Sci Technol 44:219–227
Pekakis PA, Xekoukoulotakis NP, Mantzavinos D (2006) Treatment of textile dyehouse wastewater by TiO_2 photocatalysis. Water Res 40:1276–1286
Peng F, Zhu H, Wang H (2007) Preparation of Ag-sensitized ZnO and its photocatalytic performance under simulated solar light. Korean J Chem Eng 24(6):1022–1026
Pierson HO (1999) Handbook of chemical vapor deposition (CVD). Principles, technology, and applications. Noyes Publication, New Jersey
Plantard G, Correia F, Goetz V (2011) Kinetic and efficiency of TiO_2-coated on foam or tissue and TiO_2-suspension in a photocatalytic reactor applied to the degradation of the 2,4-dichlorophenol. J Photochem Photobiol A 222:111–116
Pujara K, Kamble SP, Pangarkar VG et al (2007) Photocatalytic degradation of phenol-4-sulfonic acid using an artificial UV/TiO_2 system in a slurry bubble column reactor. Ind Eng Chem Res 46:4257–4264
Puma GL, Yue PL (1999) Enhanced photocatalysis in a pilot laminar falling film slurry reactor. Ind Eng Chem Res 38:3246–3254
Qin W, Zhang D, Zhao D (2010) Near-infrared photocatalysis based on YF_3: Yb^{3+},Tm^{3+}/TiO_2 core/shell nanoparticles. Chem Commun 46:2304–2306
Randeniya LK, Bendavid A, Martin PJ et al (2007) Photoelectrochemical and structural properties of TiO_2 and N-Doped TiO_2 thin films synthesized using pulsed direct current plasma-activated chemical vapor deposition. J Phys Chem C 111:18334–18340

Ranjit KT, Viswanathan VJ (1997) Synthesis, characterisation and photocatalytic properties of iron-doped TiO_2 catalysts. J Photochem Photobiol A 108:79–84

Ray AK, Beenackers AACM (1998) Novel photocatalytic reactor for water purification. AIChE J 44:477–483

Riegel G, Bolton JR (1995) Photocatalytic Efficiency Variability in TiO_2 Particles. J Phys Chem 99:4215–4224

Robertson PKJ, Bahnemann DW, Robertson JMC (2005) Photocatalytic detoxification of water and air. In: Boule P, Bahnemann DW, Robertson PKJ (ed) Hdb Env Chem, vol 2. Springer-Verlag, Heidelberg, pp 367–423

Rouquerol F, Rouquerol J, Singh K (1999) Adsorption by powders and porous solids principles. Academic Press, London

Sampath S, Uchida H, Yoneyama H (1994) Photocatalytic decomposition of gaseous pyridine over zeolite-supported titanium dioxide. J Catal 149:189–194

Saquib M, Muneer M (2003) TiO_2-mediated photocatalytic degradation of a triphenylmethane dye (gentian violet), in aqueous suspensions. Dyes Pigm 56:37–49

Sasikala R, Shirole AR, Sudarsan V et al (2010) Enhanced photocatalytic activity of indium and nitrogen co-doped TiO_2–Pd nanocomposites for hydrogen generation. Appl Catal A 377:47–54

Sclafani A, Herrmann JM (1996) Comparison of the photoelectronic and photocatalytic activities of various anatase and rutile forms of titania in pure liquid organic phases and in aqueous solutions. J Phys Chem 100:13655–13661

Shao H-F, Qian X-F, Zhu Z-K (2005) The synthesis of ZnS hollow nanospheres with nanoporous shell. J Solid State Chem 178:3522–3528

Shchukin DG, Sukhorukov GB (2004) Nanoparticles synthesis in engineered organic nanoscale reactors. Adv Mater 16:671–682

Shchukin D, Ustinovich E, Sviridov D et al (2004) Effect of silver deposits on the photocatalytic activity of titanium dioxide for the removal of 2-chlorophenol in water. Photochem Photobiol Sci 3:142–144

Shen S, Guo L (2008) Growth of quantum-confined CdS nanoparticles inside Ti-MCM-41 as a visible light photocatalyst. Mater Res Bull 43:437–446

Shourong Z, Qingguo H, Jun Z (1997) A study on the photoremoval of TiO_2 suspension solution. J Photochem Photobiol A 108:235–238

Sivalingam G, Nagaveni K, Hegde MS et al (2003) Photocatalytic degradation of various dyes by combustion synthesized nano anatase TiO_2. Appl Catal B 45:23–38

Škapin SD, Dražič G, Orel ZC (2007) Microstructure of nanoscale zinc oxide crystallites. Mater Lett 61:2783–2788

Slavchov R, Ivanov T, Radoev B (2007) Screened potential of a charged step defect on a semiconductor surface. J Phys Condens Matter 19:1–6

Su C, Hong B-Y, Tseng C-M (2004) Sol–gel preparation and photocatalysis of titanium dioxide. Catal Today 96:119–126

Sudhagar P, Sathyamoorthy R, Chandramohan S (2008) Influence of porous morphology on optical dispersion properties of template free mesoporous titanium dioxide (TiO_2) films. Appl Surf Sci 254:1919–1928

Sun H, Wang C, Pang S et al (2008) Photocatalytic TiO_2 films prepared by chemical vapor deposition at atmosphere pressure. J Non-Cryst Solids 354:1440–1443

Surolia PK, Tayade RJ, Jasra RV (2007) Effect of anions on the photocatalytic activity of Fe(III) salts impregnated TiO_2. Ind Eng Chem Res 46:6196–6203

Tachibana Y, Moser JE, Grätzel M (1996) Subpicosecond interfacial charge separation in dye-sensitized nanocrystalline titanium dioxide films. J Phys Chem 100:20056–20062

Ullah R, Dutta J (2008) Photocatalytic degradation of organic dyes with manganese-doped ZnO nanoparticles. J Hazard Mater 156:194–200

Uzunova-Bujnova M, Todorovska R, Dimitrov D (2008) Lanthanide-doped titanium dioxide layers as photocatalysts. Appl Surf Sci 254:7296–7302

Vohra MS, Tanaka K (2001) Enhanced photocatalytic activity of nafion-coated TiO_2. Environ Sci Technol 35:411–415

Wang Y, Cheng H, Zhang L (2000) The preparation, characterization, photoelectrochemical and photocatalytic properties of lanthanide metal-ion-doped TiO_2 nanoparticles. J Mol Catal A Chem 151:205–216

Wang K-H, Jehng J-M, Hsieh Y-H et al (2002) The reaction pathway for the heterogeneous photocatalysis of trichloroethylene in gas phase. J Hazard Mater 90:63–75

Wang R, Xin JH, Yang Y et al (2004) The characteristics and photocatalytic activities of silver doped ZnO nanocrystallites. Appl Surf Sci 227:312–317

Wang Y, Li X, Wang N et al (2008) Controllable synthesis of ZnO nanoflowers and their morphology-dependent photocatalytic activities. Sep Purif Technol 62:727–732

Ward DK, Ko EI (1996) Preparing Catalytic Materials by the Sol-Gel Method. Ind Eng Chem Res 34:421–433

Wei H, Wu Y, Lun N et al (2004) Preparation and photocatalysis of TiO_2 nanoparticles co-doped with nitrogen and lanthanum. J Mater Sci 39:1305–1308

West RH, Celnik MS, Inderwildi OR et al (2007) Toward a comprehensive model of the synthesis of TiO_2 particles from $TiCl_4$. Ind Eng Chem Res 46:6147–6156

Whittingham MS (1996) Hydrothermal synthesis of transition metal oxides under mild conditions. Curr Opin Solid State Mater Sci 1:227–232

Wu M, Long J, Huang A et al (1999) Microemulsion-mediated hydrothermal synthesis and characterization of nanosize rutile and anatase particles. Langmuir 15:8822–8825

Wu M, Lin G, Chen D et al (2002) Sol-hydrothermal synthesis and hydrothermally structural evolution of nanocrystal titanium dioxide. Chem Mater 14:1974–1980

Xiao Q, Zhang J, Xiao C et al (2007) Photocatalytic decolorization of methylene blue over $Zn_{1-x}Co_xO$ under visible light irradiation. Mater Sci Eng B 142:121–125

Xiao Q, Si Z, Zhang J (2008) Photoinduced hydroxyl radical and photocatalytic activity of samarium-doped TiO_2 nanocrystalline. J Hazard Mater 150:62–67

Xie R-C, Shang JK (2007) Morphological control in solvothermal synthesis of titanium oxide. J Mater Sci 42:6583–6589

Yang J, Mei S, Ferreira JMF (2000) Hydrothermal synthesis of nanosized titania powders: influence of peptization and peptizing agents on the crystalline phases and phase transitions. J Am Ceram Soc 83(6):1361–68

Yang J, Chen C, Ji H et al (2005) Mechanism of TiO_2-Assisted photocatalytic degradation of dyes under visible irradiation: photoelectrocatalytic study by TiO_2-film electrodes. J Phys Chem B 109:21900–21907

Yang HG, Liu G, Qiao SZ et al (2009) Solvothermal synthesis and photoreactivity of anatase TiO_2 nanosheets with dominant {001} facets. J Am Chem Soc 131:4078–4083

Yeber MC, Rodríguez J, Freer J et al (2000) Photocatalytic degradation of cellulose bleaching effluent by supported TiO_2 and ZnO. Chemosphere 41:1193–1197

Yu J, Yu X (2008) Hydrothermal synthesis and photocatalytic activity of zinc oxide hollow spheres. Environ Sci Technol 42:4902–4907

Yu J, Zhang L, Cheng B et al (2007) Hydrothermal preparation and photocatalytic activity of hierarchically sponge-like macro-/mesoporous titania. J Phys Chem C 111:10582–10589

Yuzawa H, Yoshida T, Yoshida H (2012) Gold nanoparticles on titanium oxide effective for photocatalytic hydrogen formation under visible light. Appl Catal B 115–116:294–302

Zainal Z, Hui LK, Hussein MZ et al (2009) Characterization of TiO_2–Chitosan/Glass photocatalyst for the removal of a monoazo dye via photodegradation–adsorption process. J Hazard Mater 164:138–145

Zainudin NF, Abdullah AZ, Mohamed AR (2010) Characteristics of supported nano-TiO_2/ZSM-5/silica gel (SNTZS): photocatalytic degradation of phenol. J Hazard Mater 174:299–306

Zhang Y, Mu J (2007) One-pot synthesis, photoluminescence, and photocatalysis of Ag/ZnO composites. J Colloid Interface Sci 309:478–484

Zhang Z, Wang C-C, Zakaria R (1998) Role of particle size in nanocrystalline TiO_2-based photocatalysts. J Phys Chem B 102:10871–10878

Zhang F, Chen J, Guan N et al (2004) Synthesis of titania-supported platinum catalyst: the effect of pH on morphology control and valence state during photodeposition. Langmuir 20:9329–9334
Zhang M, Sheng G, Fu J et al (2005) Novel preparation of nanosized ZnO–SnO_2 with high photocatalytic activity by homogeneous co-precipitation method. Mater Lett 59:3641–3644
Zhang L, Cheng H, Zong R et al (2009a) Photocorrosion suppression of ZnO nanoparticles via hybridization with graphite-like carbon and enhanced photocatalytic activity. J Phys Chem C 113:2368–2374
Zhang H, Chen G, Bahnemann DW (2009b) Photoelectrocatalytic materials for environmental applications. J Mater Chem 19:5089–5121
Zheng Y, Zheng L, Zhan Y (2007) Ag/ZnO Heterostructure nanocrystals: synthesis, characterization, and photocatalysis. Inorg Chem 46:6980–6986
Zhiyong Y, Mielczarski E, Mielczarski J et al (2007) Preparation, stabilization and characterization of TiO_2 on thin polyethylene films (LDPE). Photocatalytic applications. Water Res 41:862–874
Zou G, Li H, Zhang Y et al (2006) Solvothermal/hydrothermal route to semiconductor nanowires. Nanotechnology 17:313–320
Zurmühl C, Popescu R, Gerthsen D et al (2011) Microemulsion-based synthesis of nanoscale TiO_2 hollow spheres. Solid State Sci 13:1505–1509

Chapter 5
Perspectives and Advances in Photocatalysis

Abstract Interest in heterogeneous photocatalysis has heightened in the past 40 years, underscoring several perspectives and advances. A collection of applications including solar to fuel cells, intelligent ink and remote photocatalysis have been closely examined. As an emerging panacea for the treatment of recalcitrant pollutants, heterogeneous photocatalytic degradation has been highlighted with specific reference to synthetic aliphatic and aromatic organic substances. What's more, various treatment technologies for integration with photocatalytic degradation have been outlined.

5.1 Background Sketch

In the recent past, there has been phenomenal expansion in the range of investigations conducted on photocatalysis. As a result, a number of topics have emerged which equally led to several research perspectives. Semiconductor photocatalysis emerged at the time attention has been paid most globally to the eradication of hazardous materials from wastewaters, synthesis of fine chemicals and the generation of alternative energy. Currently, photocatalysis is effective for control of recalcitrant organic compounds in both water and air streams (Jacoby et al. 1996; Zuo et al. 2006). This technology is also being extensively investigated as potential for organic synthesis (see Chap. 6). At present, the photocatalytic elimination of air-borne pollutants is garnering interest and many novel catalysts are continually being tested for performance assessment (Zhang et al. 2006a). Accordingly, semiconductor photocatalysts have been tailored to result in enhanced photodegradation, or selectivity for organic synthesis (Muradov et al. 1996). There are myriad studies on common pollutants from both organic and inorganic class, but there are more studies on organic pollutants as one would expect since most pollutants belong to this class. In this chapter we give an update on the developments in the photocatalytic removal of important groups of persistent organic pollutants that may be present in air and water. We sample and discuss herein the findings of various workers regarding reactivity of both the photocatalyst and the pollutant as they affect degradation, mineralisation and their mechanism.

Conventional methods of treatment of pollutants have been scored low for many reasons which include non-destructiveness (Abramović et al. 2004), slow degradation rate at high contaminant levels (Pelizzetti et al. 1990) and ineffectiveness at low

U. I. Gaya, *Heterogeneous Photocatalysis Using Inorganic Semiconductor Solids*, DOI 10.1007/978-94-007-7775-0_5,

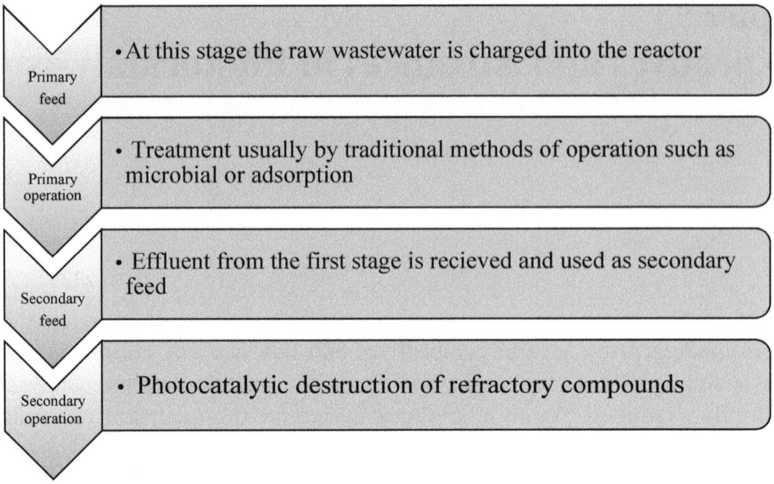

Fig. 5.1 Workflow in sequential treatment methods

pollutant concentrations. These developments paved way for advanced oxidation processes. Advanced oxidation technologies constitute a collection of established treatment methods which rely on the photogeneration of highly oxidising radical species to effect the green destruction of persistent pollutants. A growing number of novel advanced treatment technologies have already been established. Among these treatment technologies semiconductor-mediated heterogeneous photocatalysis has gained special interest owing to variety of attributes and the potential to gradually replace all conventional treatment methods. Research commitments are still made to the future of conventional methods by integrating these methods with photocatalysis in order to ensure total removal of pollutants. Photocatalysis may be operated subsequent to these methods lest treatment plants continue to be challenged with handling effluents containing high volumes of solid, refractory, synthetic chemical compounds. For example, in adsorption-photocatalysis the pollutants are first removed by using an efficient adsorbent then passed through photocatalysis. Similarly, biodegradation may be designed to precede photocatalytic operation in order to remove biodegradable contaminant first then later achieve total removal of the refractory pollutants by photocatalysis.

Generally, photocatalysis may be combined with biological, chemical or physical operation (Augugliaro et al. 2006). Physical processes that may be coupled with photocatalysis include electric bias or sonication, usually termed photoelectrocatalysis and sonophotocatalysis, respectively. Chemical processes may be ozonation, hydroperoxidisation and the addition of electron scavenging anions which are well known to enhance photodegradation rate. In this chapter, we categorise these methods according to the mode of operation as either simultaneous or sequential treatment methods. Therefore, either single reactor or a series of reactors are required for a given operation. Typical steps involved in sequential treatment are shown in Fig. 5.1.

The workflow shown in Fig. 5.1 is self explanatory. Basically, the sequential treatment methods consist of at least two main operations. The choice of the method to operate first depends on the type of pollutants. Usually for effective treatment, the conventional treatment method is usually applied as the primary operation while photocatalysis is the secondary operation. As will be discussed in this chapter, the reverse arrangement will usually result in poor mineralisation.

5.2 Progress in the Photocatalytic Degradation of Organic Compounds

5.2.1 Pesticides

Pesticides are used as in Agriculture as fungicides, insectides, herbicides and miticides. Annually, there are 3 million acute pesticide-poisoning with about 220,000 deaths (Ojanperä 2000). From remediation standpoint, pesticides have been classed as persistent organic pollutants due to their resistance to natural degradation processes and ability to persist in the environment for long periods of time. Isoproturon, atrazine (**1**), 2,4-dichlorophenoxyacetic acid (an organochlorine pesticide) (**2**) and paraquat (**3**) have been named among the top ten pesticides used in the world (Barceló and Hennion 2008). Most of these pesticides have widespread use in agriculture as herbicides thereby risking the aquatic and terrestrial environment.

As demonstrated by several studies, photocatalytic oxidation is an effective and suitable process for degradation of pesticides from different organic classes (L'homme et al. 2008). In addition, heterogeneous photocatalysis is effective irrespective of the heterogeneous phase. For instance, Sleiman et al. (2008) investigated for the first time the TiO_2 photocatalysis for the removal of pesticides in the gas phase. Dichlorvos (**4**) (an organophosphate pesticide) was used by these workers as a model pesticide, and experiments were carried out using both static and dynamic reaction systems to explore the different aspects of the process. The results showed an immediate, total degradation of dichlorvos at ppbv levels (50–350 ppbv) as well as 50–85 % mineralisation. The extent of mineralisation and the formation of reaction intermediates were significantly affected by relative humidity. Mineralisation was improved from 50 to about 60 % by loading activated carbon. Continuous irradiation for a week showed no sign of deactivation of the carbon-loaded catalyst. The photoactivity enhancement was attributed partly to the corresponding enhancement of adsorption of pollutants. Konstantinou et al. (2001) has reported the degradation of propanil and molinate in aqueous solutions containing TiO_2 suspensions under simulated solar irradiation. The photocatalytic degradation of propanil and molinate was a fast process with half-lives varying from 4.3 to 2.9 min respectively, and followed pseudo-first-order kinetics based on the Langmuir–Hinshelwood model. The mineralization of the organic carbon to CO_2 after 240 min of irradiation was found to be remarkable ($\geq$ 95 %) for both herbicides.

The solar photodegradation of sevnol, a commercial pesticide, was investigated by García et al. (2006) based on the concentration of carbaryl as its active part. Experiments were carried out both at laboratory and pilot plant scale using TiO_2 as catalyst. Complete dissappearance of carbaryl was achieved, while total mineralisation required longer irradiation. The results obtained with commercial sevnol were consistent with those of pure carbaryl, although the reaction was slower. Other pesticides degraded by photocatalysis include 2,4-dichlorophenoxyacetic acid (**2**) and 3-(3,4-dichlorophenyl)-1,1-dimethylurea or diuron (**5**) (a phenylurea pesticide) (Malato et al. 2003; Kamble et al. 2004).

(1)

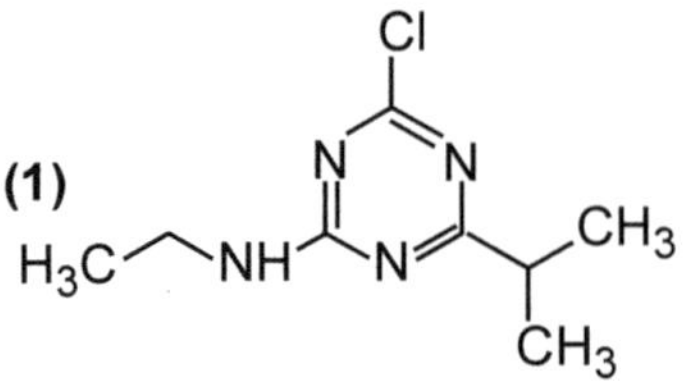

4-chloro-*N*-ethyl-6-isopropyl-1,3,5-triazin-2-amine (atrazine)

(2)

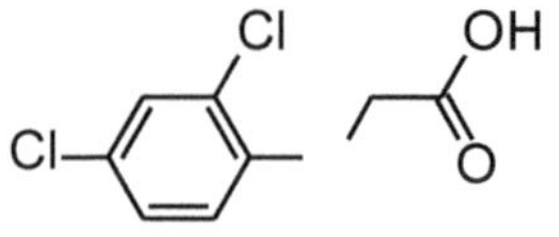

(2,4-dichlorophenoxy)acetic acid (2,4-D)

(3)

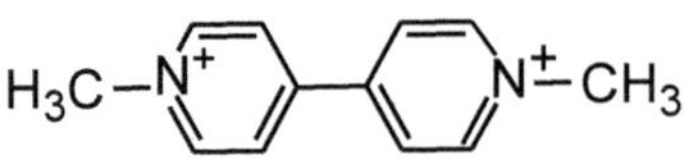

1,1'-dimethyl-4,4'-bipyridinium (paraquat)

(4)

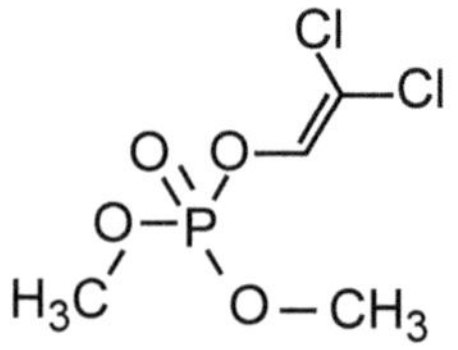

2,2-dichlorovinyl dimethyl phosphate (dichlorvos)

(5)

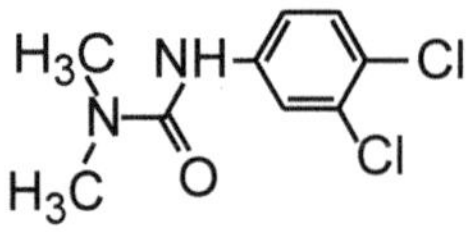

3-(3,4-dichlorophenyl)-1,1-dimethylurea (diuron)

(6)

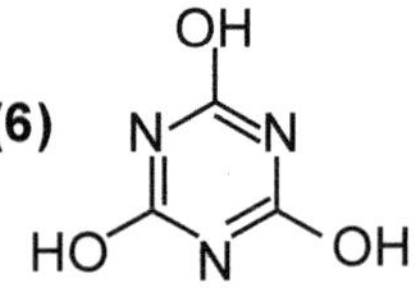

1,3,5-triazine-2,4,6-triol (cyanuric acid)

(7)

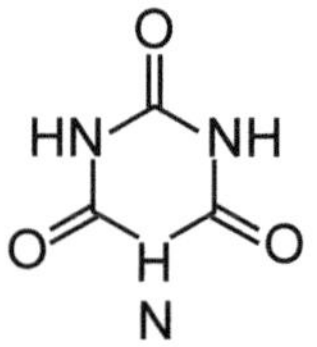

1,3,5-triazinane-2,4,6-trione(isocyanuric acid)

Even though the effect of different TiO_2 samples has been extensively studied, no major effort has been made to compare effect of −Cl on the degradability of organochlorine pesticides. It appears that organochlorine pesticides are more susceptible to photocatalytic degradation than non-chlorinated organic compounds. This observation is be supported by the following findings. Qamar et al. (2006) have observed faster photoctalytic degradation of triclopyr on TiO_2 than daminozid. Similarly, dichlorvos which contains two chlorine atoms degrades faster than phosphamidon; and Degussa P25 TiO_2 was more efficient than other TiO_2 samples (Rahman and Muneer 2005). But it may be inferred also from the aforementioned that the higher the number of chlorines the more degradable the pesticide. However, this conclusion may not be the case when fully chlorinated rings are encountered. A comparison of the photocatalytic oxidation has shown relatively more stability of lindane over phenol and chlorinated phenols, which are known intermediaries of pesticide synthesis (Dionysiou et al. 2000). The allicyclic ring in lindane is fully chlorinated which would perhaps result in difficulty in the scission of ring-Cl bond and subsequent decomposition of the remaining core structure. Photocatalytic degradation of pesticides can be promoted by the addition of certain enhancers such as surfactants. It has been shown that pesticide adsorption and subsequent degradation can be enhanced by the presence of cetyltrimethyl ammonium bromide (Zhu et al. 2007). The addition of H_2O_2 was also found to affect the photodestruction rate positively.

One of the pesticide group that received much interest in environmental photocatalysis is *s*-atrazines as these degrade fast but usually yield intermediates that transform into a highly stable and resistant cyanuric acid (2,4,6-trihydroxy-1,3,5-triazine) (**6**), a mild toxic compound. Even though atrazine is a banned or regulated substance in several countries, 100 ngl^{-1} to 1 μgl^{-1} levels have been reported in surface waters sampled across the world (Parra et al. 2004). Atrazine could not be mineralised under sonolysis (20 kHz); ozonation; photolysis; TiO_2 photocatalysis, employed either separately or in combination (McMurray et al. 2006). Pelizzetti et al. (1996) was perhaps the first to observe the rapid photocatalytic degradation of atrazine herbicide and its final defiance to full mineralisation. An investigation has shown that atrazine can be degraded by direct photolysis without being mineralised (Bianchi et al. 2006). Acceleration of degradation and part mineralisation to less than half of the initial amount can be realised by the addition of small amount of semiconductor photocatalyst such as TiO_2. Both degradation and mineralisation processes can be further improved by the addition of peroxodisulphate. In support of these results, Lackhoff and Niessner (2002) showed that semiconductor photocatalysts such as TiO_2 and ZnO maintain the expected atrazine photocatalytic degradation rates which are significantly faster than direct photolysis, with rate constants in this case $27–327 \times 10^{-3}$ min^{-1} and 0.10×10^{-3} min^{-1}, respectively. To date, cyanuric acid remains the only organic compound persistent against conventional photocatalytic mineralisation (Minero et al. 1996).

Recently, a few studies show evidence of mineralisation of atrazine. Horikoshi and Hidaka (2003) have shown complete mineralisation of atrazine monitored based on nitrogen and chlorine disappearance in presence of UV-illuminated TiO_2, under hydrothermal or supercritical aqueous condition, although these processes may not be industrially applicable and cannot be established as really photocatalytic based to the extreme conditions applied (23 MPa at 350 °C or 410 °C vis-à-vis 120 cm^3 reactor). Other workers reported almost total mineralisation (97 %) of atrazine in electro-fenton process mediated by boron-doped diamond (Oturan et al. 2012). In a different study, considerable cyanuric acid was degraded (75 % in 300 min) by singlet oxygen generated on F-TiO_2 (Jańczyk et al. 2006). Predictably, combined methods such as electrophotocatalysis can be thought of as a potential method for the total mineralisation of atrazine.

The chemical properties of cyanuric acid may have contributed significantly to its resistance to photocatalytic degradation. Cyanuric acid is a stable substrate, poorly soluble in water, and is a weak acid in aqueous media ($pK_{a1} = 7.0$, $pK_{a2} = 11.3$, and $pK_{a3} = 14.5$), which occurs as enol in a tautomeric mixture with 2,4,6-trioxo-s-triazine or isocyanuric acid (**7**) (the keto form) (Jańczyk et al. 2006; García-López et al. 2007). In alkaline solution, the acid exists in the enolic form, whereas in acidic media ($pH < 6$), the keto form is the more stable tautomer.

Often the resistance of cyanuric acid has been attributed to the rapid and reversible addition of aromatic nucleus. However, this opinion has been refuted based on primary kinetic isotopic effect. Trimethyl cyanurate shows kinetic isotope effect ($k_H/k_D = 3.1$–3.3) on degradation selectivity when the methyl groups are deuterated, and cyanuric acid enriched in ^{18}O does not exchange hydroxyl groups with bulk water under these conditions (Tetzlaff and Jenks 1999). The former indicates hydrogen abstraction while the latter rules out the possibility of reversible addition reaction such as tautomerism, from and to the original cyanuric acid. While expecting further explanation on the recalcitrance of cyanuric acid in photocatalytic solutions we may at present simply say this compound is resistant to photo-oxidant's attack.

5.2.2 Benzene, Toluene, Xylene (BTX)

The primary sources of BTX are refinery streams, especially from catalytic reforming and cracking, and pyrolysis gasoline from steam cracking and from coal liquids (Matar and Hatch 2000). Therefore, air can be contaminated by emissions from these refinery units. The direct and selective hydroxylation of benzene (among volatile organic compounds) to phenol has been one of the difficult and challenging goals in both heterogeneous and homogeneous catalysis as a potentially alternative route of the cumene process. This perhaps invited intensive research into the photo-oxidation of benzene (Fujishima et al. 2001; Zama et al. 2000). In attempt to investigate the kinetic model and degradation mechanisms, the photocatalytic oxidation of benzene (1 $mgdm^{-3}$) in air in a batch reactor containing 0.2 g of TiO_2/Sr_2CeO_4 catalysts was carried out by Zhong et al. (2007). The kinetic data matched well with the Langmuir–Hinshelwood kinetic model with the limiting rate constant and the adsorption constant of 0.0064 $mgdm^{-3}\ min^{-1}$ and 9.2078/mg^{-1}, respectively. No gas phase intermediates were detected by direct Gas Chromatography using Flame Ionisation Detector (GC/FID) analysis under the conditions of the study despite the high benzene concentration. However, GC-MS analysis of ether-extracted surface species confirmed the presence of (3-methyl-oxiran-2-yl)-methanol and ethyl acetate and detected small concentrations of 2,6-bis(1,1-dimethylethyl)-4,4-dimethylycyclohe, 2,5-cyclohexadiene-1,4,dione, 2,6-bis(1,1-dim) and butylated hydroxytoluene.

The presence of La in TiO_2 favours cleavage of benzene ring. In an investigation of the photocatalytic degradation benzene on sol gel La-TiO_2 coated glass, first order kinetics was observed but CO and CO_2 formation was consistent with zero order kinetic scheme (Zhang et al. 2006b). The intermediates of benzene transformation on pure TiO_2 film were phenol, hydroquinone and 1,4-benzoquinone while on La-doped TiO_2 2-methyl-crotonaldehyde, 2-hexanol and 4-hydroxyl-3-methyl-2-butanone which confirms that La-TiO_2 assists ring cleavage. Conversely, surface fluorination is known to affect negatively the degradation of benzene, toluene or xylene (Kim and Choi 2007).

Studies prove the possibility of achieving the complete mineralization of benzene by using the platinized titania catalyst and optimizing such parameters as the reaction temperature, space time, and the concentrations of oxygen and water vapor in the feed stream. The photocatalytic degradation of benzene in oxygen-containing gaseous feed streams was investigated in a fixed-bed annular photoreactor packed with 19 g of sol-gel synthesised titania (Fu et al. 1995). Humidity was 50 %. The workers observed rapid oxidation of benzene to carbon dioxide and water without forming any detectable organic reaction products in the reactor effluent, although only some of the benzene reacted. Both the overall conversion of benzene and its mineralization were improved by platinising the titania with as low as 0.1 wt%. This approach improved both photocatalytic and thermocatalytic reactions. Rates of photocatalytic reactions were significantly enhanced between 70 and 90 °C while at temperatures above 90 °C the rates of thermocatalytic oxidation reactions were noticeably increased. Later on, an independent investigation of photocatalytic oxidation of benzene in gas phase was carried out by Einaga et al. (2001) with a flow reactor in a humidified airstream (2.2 %) at room temperature. Benzene was quantitatively decomposed to CO_2 in presence of UV-irradiated 1.0 wt%-Pt/TiO_2 catalyst. The findings of these workers observed that the photooxidation of benzene to CO_2 over Pt/TiO_2 catalyst proceeds by two sequential steps. Firstly, benzene is decomposed to CO_2 and CO with the selectivities of 94 % and 6 % in that particular study, respectively. Secondly, CO is subsequently oxidized to CO_2.

The photo-oxidation of toluene has been carried out over polycrystalline anatase TiO_2 by Augugliaro et al. (1999). The reactor used was a pyrex fixed-bed continuous photoreactor and the reacting gaseous mixture was toluene, air and water in various molar ratios under irradiation by a medium pressure Hg lamp. The presence of oxygen was essential for the occurrence of the photoreaction while water played an important role in order to maintain the activity of TiO_2. Benzaldehyde was the major product, but benzene, benzyl alcohol and traces of benzoic acid, phenol and other unidentified compounds were detected in the course of degradation. The molar conversion to benzaldehyde with respect to the initial amount of toluene was ca. 20 % under the best experimental conditions. Enhancement of the photocatalytic destruction of toluene was observed by mixing feeds with halomethane (CH_2Cl_2, $CHCl_3$), 1,1,3-trichloropropene (Sauer et al. 1995) and trichloroethene (Luo and Ollis 1996). Tetrachloromethane (CCl_4) was a poor promoter and does not degrade appreciably. Some many workers have observed catalyst deactivation during toluene vapour degradation studies which has often been attributed to the presence of benzoic acid, benzaldehyde and scarcity of water (Sauer et al. 1995; Luo and Ollis 1996; Martra et al. 1999; Maira et al. 2001; Blount and Falconer 2002; Demeestere et al. 2008). This calls for the synthesis of a more suitable catalyst for toluene removal.

5.2.3 *Pharmaceuticals*

Pharmaceuticals and personal care products (PPCPs) are a new class of pollutants (Li et al. 2011). Pharmaceutical residues may pose great challenge during deactivation in sewage treatment plants. Pharmaceutical compounds have the tendency to cause the development of antibiotic resistant microbes in the aquatic environment (Dalrymple et al. 2007). Unwanted therapeutic medicines may be disposed improperly in our surroundings while unmetabolised ones in urine or excreta may reach sewage treatment plants in substantial quantities (Castiglioni et al. 2006). Removing these pharmaceuticals is more challenging than preventing their widespread contamination due to the poor efficiency of the many treatment methods and cost. Even though the environmental concentrations of pharmaceuticals are generally at trace levels, typically within $ngdm^{-3}$ to μgdm^{-3}, but these levels can be sufficient to induce toxic effects (Hernando et al. 2006).

Complete mineralisation of pharmaceuticals of therapeutic class (such as tetracycline) in aqueous TiO_2 suspensions has been reported (Reyes et al. 2006). Even though photolysis can serve the degradation of pharmaceuticals this process is very slow compared to photocatalysis. For instance, Calza et al. (2008) has observed fast and complete degradation of amiloride (within 30 min of irradiation) by heterogeneous photocatalysis. Mineralisation was however found to be slow as 25 % of total organic carbon (TOC) remained in solution after 240 min. The same compound took 240 min to degrade by homogeneous photolysis. The tolerance of amiloride against photocatalytic mineralisation was related to the stability of guanidine, a major intermediate accounting for 17 % of the residual TOC. The tendency of certain pharmaceuticals to be resistant and to leave behind degradation residues is evident from many researches. For example, paracetamol took 450 min to yield 85 % disappearance of the initial compound from irradiated aqueous TiO_2 solutions (Yang et al. 2009). Because their photocatalytic degradation may not correspond to their mineralisation, pharmaceutical wastes must therefore be handled with care.

(8)

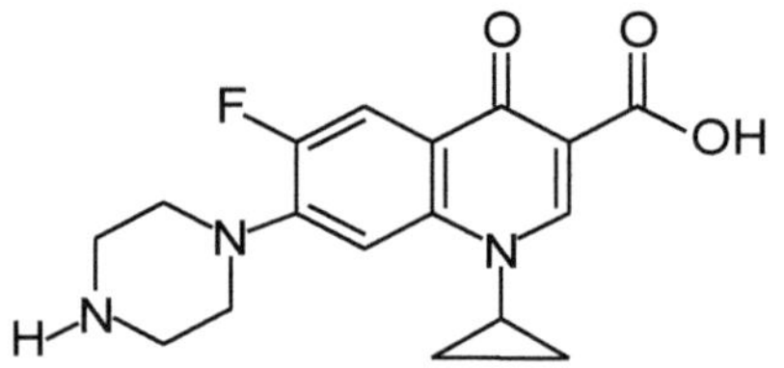

1-cyclopropyl-6-fluoro-4-oxo-7-piperazin-1-yl-1,4-dihydroquinoline-3-carboxylic acid (ciprofloxacin)

(9)

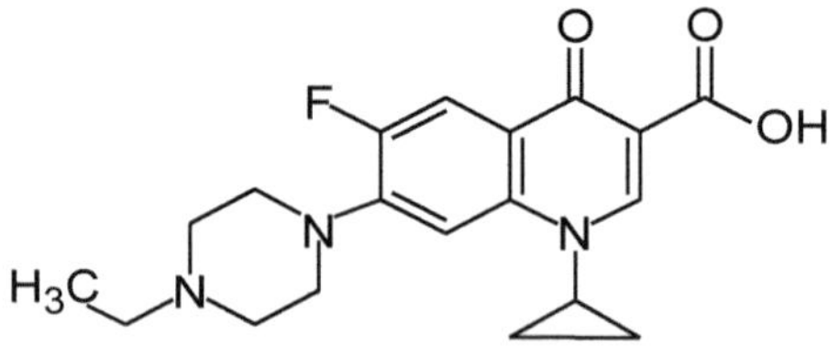

1-cyclopropyl-7-(4-ethylpiperazin-1-yl)-6-fluoro-4-oxo-1,4-dihydroquinoline-3-carboxylic acid (enrofloxacin)

(10)

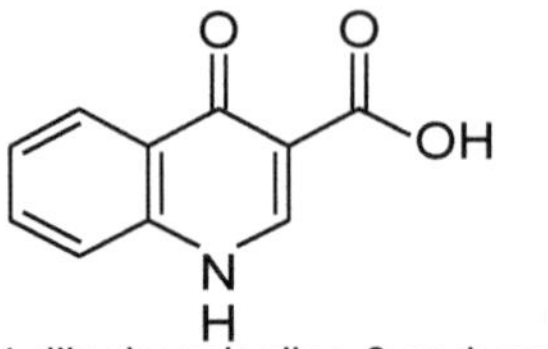

4-oxo-1,4-dihydroquinoline-3-carboxylic acid (oxohydroquinolinecarboxylic acid nucleus)

(11)

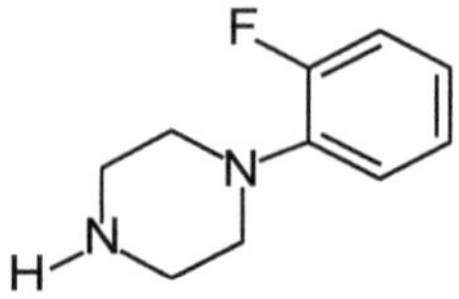

1-(2-fluorophenyl)piperazine

It may be appropriate to arguably state that the efficiency of photocatalytic removal of pharmaceuticals as well as other pollutants depends in the main upon the vincibility of the pollutant to photocatalytic oxidation, form of irradiation, photoactivity and type of the photocatalyst. For example, Chatzitakis et al. (2008) achieved 93 % mineralisation of chloramphenicol after 180 min of illumination in the presence of TiO_2 P-25 (anatase/rutile ratio = 3.6) while in the presence of TiO_2 (A) (100 % anatase) only 60 % mineralisation was reached. Similarly, Paul et al. (2007) has studied the mechanism, kinetics, and transformation of byproducts of fluoroquinolone antibacterial agents (ciprofloxacin (**8**), enrofloxacin (**9**), norfloxacin, and flumequine) via visible-light TiO_2 photocatalysis. The results of the study were compared with those of UV photocatalysis under the same conditions. These workers observed the occurrence of visible-light photocatalytic degradation of ciprofloxacin and three structural analogues (flumequine, enrofloxacin, and norfloxacin) that retain the oxo-quinolinecarboxylic acid core structure (**10**) while in contrast, no visible-light degradation was observed for 1-(2-fluorophenyl)piperazine (**11**), an analogue that lacks this core structure. The visible pathway produced intermediates that are thought to form through 1-electron oxidation, assisted by electron transfer to TiO_2 from previously excited surface-complexed fluoroquinolone molecule.

5.2.4 *Dyes*

Dyes constitute an important class of water pollutants with high resistance to biodegradability. Because the nature of the dye affects its reactivity we therefore consider it important to introduce the classification of dye. There are several methods of classification of dyes. The dye manufacturers and chemists adopt the classification based on chemical structure (Ali et al. 2005). By this method dyes are classified as acridine dyes (such as acridine O), azo dyes (such as congo red), arylmethane dyes (such as malachite green), anthroqinone dyes (such as carmine), nitro dyes (such as napthol), xanthenes dyes (such as rhodamine B) and quinine–amine dyes. Azo dyes are the largest group of synthetic colorants (60–70 %) and are being used in industry for applications such as textiles, papers, leathers, gasoline, additives, foodstuffs, cosmetics, laser materials, xerography and laser printing (Rauf et al. 2011). Azo-dyes contain at least one azo groups (-N = N-) on one or two aromatic rings. Examples of these azo dyes include methyl orange, reactive red 120, acid orange 7 and methyl red.

(12)

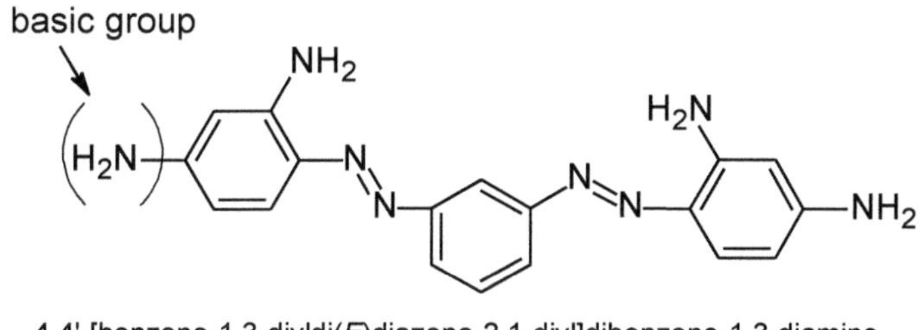

4,4'-[benzene-1,3-diyldi(*E*)diazene-2,1-diyl]dibenzene-1,3-diamine
(basic brown 1)

The second approach to classification of dyes is based on the chemical nature, method of application and properties of the dye. This method of classification is preferred by the dye users and is barely considered in photocatalysis whenever dyes are discussed. In this approach dyes are classified into various classes and we wish to discuss herein the water-soluble classes only for their commonness in the literature concerning photocatalysis. Acidic dyes are water-soluble anionic dyes containing acidic groups such as sulphonic groups, and are suitable for use in acidic solutions. Basic dyes such as basic brown 1 (**12**), are water-soluble cationic groups usually of amino and substituted amino compounds. Direct dyes such as direct blue 71 (**13**), are also water soluble and anionic but their acid group is not the means of adsorption. This class of dyes comprises of polyazo compounds, along with some stilbenes, phthalocyanines and oxazines.

(13)

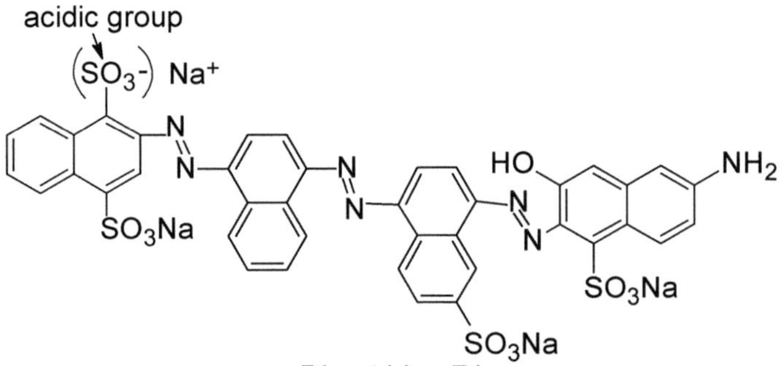

Direct blue 71

Anionic and cationic dyes have widespread use in important manufacturing industries. While anionic dyes of both acid and direct group are mostly used to dye nylon, silk and wool, modified cationic dyes are used to dye paper, polyacrylonitrile, modified nylons, modified polyesters and in making inks. Although, most dyes are facilely degraded or mineralised by heterogeneous photocatalysis, the degradation rate depends upon the family of the dye (Epling and Lin 2002) and the semiconductor photocatalyst used. A comparison of the initial rate of degradation of acid brown 14 on TiO_2, ZnO, SnO_2, ZrO_2, CdS, α-Fe_2O_3 and WO_3 showed ZnO to be the most active photocatalyst and therefore confirmed its unrivalled absorption of light quanta from the spectrum (Sakthivel et al. 2003). One important point worthmentioning is that, the semiconductor accelerated photodegradation of anionic dyes occurs more rapidly in acidic media than in neutral or alkaline media owing to the positively charged state of the semiconductor particle in the former medium and the adsorption behaviour of the dye (Zhang et al. 1998). These electrostatic behaviours as well as the efficiency of doped-ZnO materials for removal of acidic dyes have been detailed in Chap. 4 of this book.

Recently, many workers have studied the degradation of anion dyes using combined methods to exploit the possible synergy between cavitation events and photocatalytic counterpart in dye decolorisation or mineralisation. The decolorization of an azo dye, reactive red 2 using TiO_2 photocatalysis coupled with water jet cavitation elevated the degradation efficiency of the dye by about 136 %, confirming a synergistic effect between semiconductor-assisted heterogeneous photocatalysis and water jet cavitation (Wang et al. 2011). The enhancement of photocatalysis by water jet cavitation can simply attributed to the deagglomeration of catalyst particles. A dual-pathway network mechanism of sonophotocatalytic degradation of dyes was proposed by Vinu and Madras (2009) based on existing mechanisms of photocatalysis and sonocatalysis. We have herein reviewed the key elementary processes as follows:

1. Adsorption of dye:

In the first place, dyes are adsorbed on the surface of the semiconductor photocatalysts.

$$TiO_2 + \text{Dye} \rightleftharpoons TiO_2 \equiv \text{Dye} \tag{5.1}$$

A direct correlation between the rate of degradation of dyes and their coverage at the surface of titania has been established which strongly suggests that the reaction essentially occurs in the adsorbed phase at the surface of TiO_2 and not in the solution as it has been sometimes claimed (Guillard et al. 2003). Surface adsorbed dye was represented by $\equiv$ Dye.

2. Photocatalytic pathway

In the photocatalytic pathway, charge carriers are generated by the TiO_2 upon the absorption of light.

$$TiO_2 \equiv \text{Dye} + \text{UV} \rightleftharpoons TiO_2{}^* \equiv \text{Dye}\left(e^-, h^+\right) \tag{5.2}$$

Hydroxyl radicals can be formed either by oxidative process (i.e. the hole pathway) or by the reductive process (i.e. the electron pathway). In the hole pathway a positive hole is transferred to a molecule of water or hydroxyl ion.

$$TiO_2(h^+) \equiv \text{Dye} + H_2O \rightarrow TiO_2 \equiv (\text{Dye}, HO^\bullet) + H^+ \tag{5.3}$$

$$TiO_2(h^+) \equiv \text{Dye} + OH^- \rightarrow TiO_2 \equiv (\text{Dye}, HO^\bullet) \tag{5.4}$$

$$TiO_2(e^-) \equiv \text{Dye} + O_2 \rightarrow TiO_2 \equiv (\text{Dye}, O_2) \tag{5.5}$$

The electron pathway proceeds through the formation of hydroperoxyl radical and hydrogen peroxide to form hydroxyl radical.

$$TiO_2 \equiv (\text{Dye}, O_2) + H^+ \rightarrow TiO_2 \equiv (\text{Dye}, HOO^\bullet) \tag{5.6}$$

$$TiO_2 \equiv (\text{Dye}, HOO^\bullet) + H^+ \rightarrow TiO_2 \equiv (\text{Dye}, H_2O_2) \tag{5.7}$$

$$TiO_2 \equiv (\text{Dye}, H_2O_2) + e^- \rightarrow TiO_2 \equiv (\text{Dye}, HO^\bullet) + OH^- \tag{5.8}$$

3. Sonophotocatalytic pathway

In the sonocatalytic pathway charge carriers are generated on TiO_2 by the combination of catalysis and ultrasonication (US).

$$TiO_2 \equiv \text{Dye} +))) \rightleftharpoons TiO_2{}^* \equiv \text{Dye}(e^-, h^+) \tag{5.9}$$

Where))) represents ultrasonication.

In the hole pathway hydroxyl radical is formed via direct hole transfer to water or hydroxyl ion.

$$TiO_2(h^+) \equiv Dye + H_2O \rightarrow TiO_2 \equiv (Dye, HO^{\bullet}) + H^+ \tag{5.10}$$

In the sonolytic pathway reactive oxygen species are formed by cavitation.

$$H_2O +))) \rightarrow HO^{\bullet} + H^{\bullet} \tag{5.11}$$

$$O_2 \rightarrow 2O^{\bullet} \tag{5.12}$$

$$H_2O + O^{\bullet} \rightarrow 2OH^{\bullet} \tag{5.13}$$

$$OH^{\bullet} + O_2^{-\bullet} \rightarrow OH^- + O_2 \tag{5.14}$$

$$H^{\bullet} + O_2 \rightarrow HO_2 \tag{5.15}$$

$$HO_2 + H^{\bullet} \rightarrow H_2O_2 \tag{5.16}$$

$$HO_2 + HO_2 \rightarrow H_2O_2 + O_2 \tag{5.17}$$

$$H_2O_2 \rightarrow \text{product} \tag{5.18}$$

Finally, the adsorption of hydroxyl radical takes place

$$TiO_2 \equiv Dye + OH^{\bullet} \rightarrow TiO_2 \equiv (Dye, OH^{\bullet}) \tag{5.19}$$

The rate constant for the dye degradation under such a process can be evaluated by Eq. (5.20) which is based on reduction technique rather than pseudo-steady state approximation.

$$A'\left(\frac{[Dye]_o}{[Dye]_t}\right) + B'\left(\frac{[Dye]_o}{[Dye]_t}\right) = C't \tag{5.20}$$

where, A', B' and C' are coefficients each representing the adsorption of the dye onto TiO_2, desorption from TiO_2 surface, generation of surface hydroxyl species by UV and US pathways and the degradation of the dye.

Owing to the persistence of dyes in the environment, photocatalysis has been used in combination with other methods. In the photoelectrocatalytic study of the degradation of various dyes under visible light irradiation using a TiO_2 nanoparticles electrode it was found that the dyes could controllably interact with the TiO_2 surface by external bias and lead to efficient photoelectrocatalytic degradation (Yang et al. 2005b). The study also proved that both the superoxide anionic radical and the dye cationic radical are essential to the mineralisation of the dyes under visible light-induced photocatalytic conditions. The degradation of both cationic and anionic dyes at different biases, including the change in the degradation rate of the dyes and the photocurrent change with the bias potential, the degraded intermediates, the voltage induced adsorption of dyes, the accumulation of electrons in the TiO_2 electrode, the effect of various additives such as benzoquinone (BQ) and N, N-dimethyl aniline, and the formation of active oxygen species such as $O_2^{\bullet -}$ and H_2O_2 were examined by UV-visible spectroscopy, HPLC, TOC, and spin-trap ESR spectrometry. The cationic dyes (rhodamine B and malachite green) underwent efficient mineralisation at negative bias, but the N-dealkylation process predominated at positive bias under visible light irradiation. It was found that the discolorations of the sulforhodamine and alizarin red could not be substantially accelerated at either negative or positive bias. Based on the simultaneous formation of $O_2^{\bullet -}$ and dye $^{\bullet +}$ at a negative bias of − 0.6 V vs. SCE, at the electrode/electrolyte interface during degradation of cationic rhodamine B it was concluded that both superoxide and and the dye cation are essential part of the degredation process.

5.2.5 Phenolic Compounds

Phenolic compounds, also referred to as phenols, can arise ubiquitously in the environment in large concentrations from agricultural activities and some industrial discharges such as coal gasification, resin manufacturing, oil refining, coking plants, chemical synthesis, dyes, plastics, textiles, pharmaceuticals, paper mill, herbicides and fungicides production. Phenols are known carcinogens and their persistence in the environment has been acknowledged. Phenol and phenol derivatives are considered one of the most common organic water pollutants because of high toxicity, even at low concentrations (Mantilla et al. 2010). In addition, phenol exhibits the basic structure of many organic molecules and can be effectively mineralised by semiconductor photocatalysis. Owing to the aforementioned facts, phenol has been used as a model pollutant in great many photodepollution studies. Often, the photocatalytic activity of newly synthesized semiconductor photocatalysts has been evaluated using phenol as model pollutant. For example, Zainuddin et al. (2010) prepared a zeolite-supported TiO_2 prepared using nano-TiO_2:ZSM-5:silica gel:colloidal silica gel ratio of 1:0.6:0.6:1 and evaluated its performance using phenol as model pollutant. The catalyst gave about 90 % degradation of 50 $mgdm^{-3}$ phenol solution in 180 min, which means higher photocatalytic performance than that of the commercial Degussa P25 which only gave 67 % degradation. The authors attributed the higher photocatalytic activity of the modified catalyst to its large specific surface area (275.7 $m^2\,g^{-1}$), small particle size (8.1 nm), high crystal quality and low electron-hole recombination factor.

Apart from evaluation of photocatalytic activity, phenol has been used as a model pollutant for effective optimization of synthesis parameters in order to have reproducible catalysts. Chen et al. (2010) applied mixture design and response surface techniques to prepare visible light-active TiO_2 photocatalysts co-doped with sulfur, carbon, and SnO_4 using $Ti(OC_3H_7)_4$, $SnCl_4$ and thiourea, as precursors. A synergic effect induced by the dopants SnO_2 and S, C on the visible light activity of TiO_2 was quantitatively established for the first time through mixture design techniques and response surface methodologies. The empirically quartic models obtained described well the main interactive influences of the precursors on the specific surface area and visible light activity of the S, C, SnO_2-codoped TiO_2 samples. Even though the application of experimental design has been more on the process, this research paves way for further application on optimization of synthesis.

Phenol has also been used to evaluate the activity of photoelectrodes for use in photoelectrocatalysis. The application of electric bias may help in minimising the recombination of photogenerated charge carriers and consequently enhances the photocatalytic activity towards organic pollutant degradation. The activity of transparent and highly porous nanocrystalline TiO_2 electrodes for phenol degradation by heterogeneous photocatalysis was investigated by Oliveira et al. (2010). Electrochemical characterization was performed for electrodes with areas of $1.0\,cm^2$ and this revealed increase in capacitance values with irradiation. Electrodes with areas higher area ($9.0\,cm^2$) were effective in the remediation of $10\,cm^3$ aqueous phenol solution ($50\ mgdm^{-3}$). When irradiated by a solar simulator, the removal of phenolic total organic carbon was 48 % after 3 h. The efficiency was significantly enhanced in the electrochemically-assisted photocatalysis; the average mineralisation was 78 % after the same period of time. The investigators have shown that almost complete mineralisation can be achieved after 6 h with a TiO_2 electrode externally connected to a Pt counter-electrode and biased at + 0.7 or + 1.1 V using a potentiostat, or by a series connection to a solar cell. Phenol degradation exhibited pseudo first order kinetics, and application of the bias potential increased the rate constant from 0.21 to $0.47\,h^{-1}$.

Chlorinated phenolic compounds are well-known environmental pollutants owing to their high potential of phytotoxicity and zootoxicity. Most chlorophenols have a phenolic moiety carrying at least a −Cl group. These compounds have greater sorption potential than desorption which imply persistence in the aquatic environment for longer periods of time (Krijgsheld and van der Gen 1986). In aquatic environment, chlorinated compounds could derive from biotic and abiotic degradation of higher chlorophenols and various phenoxyherbicides (Brillas et al. 2000). Certain chlorophenols are likely to form as by-products of industrial processes such as water disinfection. A significant amount of 2,4,6-trichlorophenol up to $1.96\ \mu gdm^{-3}$ has been detected during the chlorination of drinking water (Nikolaou et al. 1999; Fingler and Dravenkar 1988). It is also known that chlorophenols are widely used as intermediaries for the synthesis of important pesticides and as pesticides themselves. Such pesticides or pesticide precursors may be released to the environment through uncontrolled use or improper waste disposal. For instance, despite the fact that dichlorophenols are in US EPA's list of priority pollutants which the community has right to know (US EPA/OSHA 2000; US EPA 2006) the widespread occurrence of dichlorophenols in open waters has been reported in several countries (Wegman and van Den Broek 1983; Abdullah and Nainggolan 1991; Gao et al. 2008). Similarly, alarming levels of 2,4,6-trichlorophenol were detected in several rivers from various geographical regions (Chiron et al. 2007). These reports have prompted the vigorous investigation of the decontamination of these environmentally toxic compounds by photocatalysis.

Generally, the degradation of phenolic compounds is known to yield genotoxic and carcinogenic intermediates such as catechol and hydroquinone which serious raise alarm. Jardim et al. (1997) have compared the toxicity of transient species of the degradation of pentachlorophenol, 2,4-dichlorophenol and 2,3,5-trichlorophenol with those of the initial compounds. These workers reported that the toxicity of the irradiated solution of pentachlorophenol to *Escherichia coli* could be higher than that of the starting material. In fact, even ordinary direct photolysis of frozen chlorophenols solution is known to yield more toxic intermediates (Bláha et al. 2004). Similarly, increase in the toxicity of irradiated aqueous methyl parathion solution was observed in the presence of ZnO which was attributed by Evgenidou et al. (2007) to the dissolusion of zinc oxide during the photodegradation process. However, the following points are worth noting. Firstly, whenever photocatalytic degradation is carried out at optimum conditions CO_2 and H_2O are eventually formed which are benign. Secondly, in a study of the photodegradation of cellulose and textile effluents on ZnO and TiO_2 photocatalysts it has been shown that the acute toxicity of overall photocatalytic solution is progressively reduced with irradiation time (Peralta-Zamora et al. 1998). This would clear the fear regarding the production of toxic intermediates. Certainly, there has been concern about the photodissolution of ZnO in irradiated solutions. A priori, the photodecomposition of ZnO is thermodynamically feasible and this has been observed during the degradation of phenolic compounds (Khodja et al. 2001). Some workers including in our laboratory have measured the amount of zinc ions in chlorophenol solutions under photodegradation in the presence of ZnO (Lathasree et al. 2004; Gaya et al. 2009; Gaya et al. 2010). However, the levels observed were mostly very low, even in more stringent conditions such as the photoelectrocatalysis of non-phenolic organic compounds (Thampi et al. 1983). Since zinc is an essential micronutrient that can be found in all tissues of the body and is essential for cell growth, differentiation, healthy immune system, and DNA synthesis (Sandstead 1991), trace dissolution of ZnO may make good news for water treatment plants. Besides the relatively extended spectral range of ZnO, it offers several other features for harvest in phenol removal. For example, the photocatalytic degradation of phenol by TiO_2 could be completely inhibited in the first 20–60 min in the presence of three crystal types of MnO_2 (α-, λ-, δ-) but with ZnO as photocatalyst the degradation only declined slightly (Li et al. 2008).

5.2.6 Aldehydes and Ketones

Oxygen-containing compounds typically comprise 5–10 % of the total organic compound concentration of automobile exhaust, and the aldehyde present in the greatest amount in automobile exhaust and usually the major aldehyde in ambient air, is methanal (Connel 2005). The photocatalytic degradation of gaseous acetaldehyde was studied by Kim and Choi (2007) using surface fluorinated TiO_2. Operation was carried out at UV lamp temperature of 27–30 °C. The UV source was a 200 Hg lamp. Acetaldedhyde (200 $mgdm^{-3}$ in 20 % O_2) was delivered at flow rate of 100 $cm^3 min^{-1}$ over 2 g of coated TiO_2. It was observed that the presence of fluorine on TiO_2 surface precludes adsorption to some extent, but resulted in more than two-fold increase in the first order rate constant for the generation of CO_2. As would be expected, the major intermediates of gas phase mineralisation of acetaldehyde is ethanoic acid, but formaldehyde was also produced in the process (Ohko et al. 1998).

In a different study, Huang et al. (2003) carried out the gas phase photocatalytic mineralisation of n-butanal in a packed-bed reactor containing TiO_2. The major intermediate products revealed by GC-MS were propanal, 1-propanol, acetaldehyde, and ethanol. These intermediates formed secondary products such as propyl formate, di-n-propyl ether, and 3-heptene presumably by esterification, dehydration, and reductive coupling. Aldol condensation of the vapor phase aldehydes on TiO_2 surfaces followed by cyclisation may be partially responsible for the formation of some minor by-products such as ethylbenzene. In the same year, investigation of the gaseous phase destruction of acetaldehyde, a major intermediate of mineralisation of butanal, was carried out over two series of nano-sized N-containing-ZnO (M: Fe, W) composite powders which indicated that the photodestruction of acetaldehyde can be enhanced by the addition of WO_3 and retarded by the addition of Fe_2O_3 (Li and Haneda 2003).

Only a few water-borne aldehydes were studied by heterogeneous photocatalysis. Two case studies of the photocatalytic destruction of formaline, a formaldehyde-methanol-water mixture and formaline wastes from the preservation of vertinarian physiologic samples have been investigated by Araa et al. (2004). Similarly, the study showed that at non-optimum conditions, photonic efficiencies were low which was attributed to the formation of methoxy compounds, formates and physisorbed formic acid which may have inhibited formation of hydroxyl radicals and consequently the photocatalytic degradation. The only ketonic compound that has attracted much interest of photocatalysis scientists is acetone, perhaps because of its widespread use as solvent. Vorontsov et al. (2000) investigated the photocatalytic oxidation of gaseous acetone in a flow-circulating reactor over platinised TiO_2 prepared by either photodeposition or reduction with $NaBH_4$. The reactor was operated at 40–120 °C. Platinised TiO_2 samples were more active than the bare counterpart, and the former did not demonstrate steady state deactivation, which was observed over TiO_2 at temperatures above 100 °C. The adsorption constants obtained from adsorption isotherms are significantly lower than those obtained from the rate-concentration curves. This corroborates the observation of acetone photoadsorption during photocatalytic oxidation. Later study by Choi et al. (2001) investigated the photocatalytic oxidation of acetone in air using one or four TiO_2-coated fibres in a continuous flow photoreactor. The kinetics of the acetone conversion was found to be consistent with zero-order kinetics. A steady-state conversion up to 80 % was achieved at ambient temperature and pressure and all the acetone molecules degraded was quantitatively converted to CO_2 with no intermediates detected. These workers observed no noticeable deactivation of the catalyst within a few hours' operation under the experimental conditions which may be due to the short life of the intermediates. However, the presence of water vapour reduced the reactivity due to the competitive adsorption on active surface site with acetone. While a measurable conversion of acetone was observed in the absence of O_2, increasing O_2 concentration up to 15 % effectively enhanced the conversion.

5.2.7 *Alkanes, Alkenes and Derivatives*

The development of the efficient oxidation catalyst for saturated hydrocarbons has been a challenging goal for chemists (Yamaguchi et al. 2006). This triggered interest into the photocatalytic oxidation of C_5 to C_7 alkanes. Boulamanti and Philippopoulos (2009) studied the gas phase photocatalytic oxidation of *n*-pentane, *iso*-pentane, *n*-hexane, *iso*-hexane and heptane over illuminated titanium dioxide in a continuous stirring-tank reactor. Conversions were over 90 % at ambient temperatures, for residence times from 50 to 85 s. The rate constants ranged from 1.87×10^{-7} mol $m^{-2}s^{-1}$ for pentane to 3.03×10^{-7} mol $m^{-2}s^{-1}$ for heptane while the adsorption constants ranged from 1.14 to 2.83 m^3 mol^{-1}, and the water adsorption constant was 11.2 m^3 mol^{-1}. One important finding of these workers is that the rate of the reaction is influenced by molecular and structural changes as follows:

1. The rate was increasing with increase in molecular weight and the presence of a tertiary carbon atom.
2. The rate of the reaction is higher for *iso*-isomers than for normal isomers.

As in most gaseous photocatalysis, no catalyst deactivation was observed. In addition, the only products observed were CO_2 and H_2O. A different study of the gas phase photocatalytic oxidation of *n*-heptane was reported by Liqiang et al. (2004) with a view to compare the lifetime of home-made TiO_2 and ZnO. The study showed that ZnO could almost loss activity in the gas phase photocatalytic oxidation of *n*-heptane while TiO_2 could keep most of its activity. However, in the gas phase photocatalytic oxidation of SO_2 both ZnO and TiO_2 were deactivated after five runs, which took 15 h. Based on surface photovoltage spectrum and X-ray photoelectron spectroscopy the deactivation was attributed to change from *n*-type to *p*-type surface owing to the adsorption of the oxidation products such as SO_3 on the semiconductor photocatalyst surface.

One of the oldest and informative gas phase photocatalysis studies was carried out by Pichat et al. (1979). The study was focused on the destruction of propene over the several UV illuminated oxides: TiO_2, ZrO_2, V_2O_5, ZnO, SnO_2, Sb_2O_4, CeO_2, WO_3, and a Sn-O-Sb mixed oxide. None of these oxides was active at 47 °C in the absence of UV irradiation corresponding to energy equal to or greater than its band gap. For propene, a rise in temperature had a negative effect on the photocatalytic activity. The quantum yields as well as selectivity varied with the metal oxides. Vanadium pentoxide was found to be photoinactive. In addition to water, CO_2, acetaldehyde, acrolein, acetone and, in some cases, propene oxide and traces of propanal were produced.

Trichloroethene appears to be the most widely investigated haloalkene in gas phase photocatalysis, which may be due its presence in industrial emissions. Trichloroethene is also widely used as industrial solvent. Often, photocatalytic studies have shown high rates of destruction of this pollutant and high quantum yields have been reached. However, direct sampling mass spectrometry and gas-phase fourier transform infrared spectroscopy (FTIR) revealed that there are significant quantities of by-products produced which include phosgene, dichloroacetyl chloride, carbon monoxide and molecular chlorine (Nimlos et al. 1979). Buechler et al. (1999) has demonstrated enhancement of the gas-phase photocatalytic oxidation of trichloroethene by controlled periodic illumination of TiO_2 films. The reactor was operated in two different geometric configurations, tubular and annular under kinetic- and diffusion-limited flow regimes. No enhancement on the rate or the selectivity of trichloroethene oxidation was observed at any of the periodic illumination conditions investigated in the kinetically limited flow regime. The ratio of trichloroethene oxidized to dichloroacetyl chloride produced was constant at all light and dark times investigated for kinetically limited operation. In diffusion-limited regimes, controlled periodic illumination increased the rate of oxidation and decreased the relative amount of dichloroacetic acid in the product stream. In the kinetic-limited regimes the apparent quantum yield was constant at around 150 % while in the diffusion-limited regimes the apparent quantum yield almost doubled.

The dependence of gas-phase ZnO-assisted photocatalytic destruction of trichloroethene on trichloroethene pressure has been observed by Driessen et al. (1998). At low pressures, CO_2 and CO were predominant carbon-based gas-phase products. At high trichloroethene pressures, more carbon-containing products were detected which include phosgene, and dichloroacetylchloride which the investigators attributed to the saturation of ZnO surface with adsorbed products mainly water, bidentate formate and dichloroacetate. These adsorbed photoproducts blocked photocatalyst sites so that partial oxidation products cannot undergo further oxidation on the ZnO surface. In addition to the intermediates mentioned above, carbon tetrachloride, chloroform, hexachloroethane, pentachloroethane, and tetrachloroethene were identified during gaseous trichloroethene degradation on TiO_2 surface (Hung and Marias 1997). Among these intermediates, chloroform conversion was very low (30 %). Wang et al. (2002) have corroborated the detection of many of the intermediates of trichloroethene vapour decomposition by FT-IR and neuclear magnetic resonance (NMR). Besides the presence of persistent intermediates, there are other factors that may affect negatively the photocatalytic degradation trichloroethene which include the presence of ozone (Shen and Ku 2002) and humidity (An et al. 2005). On the whole, the treatment of trichloroethene must therefore be done with care as more hazardous gases such as phosgene ($COCl_2$) or its components (CO and Cl_2) may leak out.

5.3 Hybrid Treatment Strategies

5.3.1 Additive-Enhanced Photocatalysis

Photocatalysis is an electron- and hole- transfer harvesting process and influencing any of these processes can affect the rate of photocatalytic reaction. Both electron and hole can be scavenged by the addition of suitable inorganic scavenger. In simultaneous TiO_2/inorganic additive photoprocess the main focus is on the use of electron scavengers such as neutral molecules (e.g. O_3, H_2O_2), cations (e.g. Fe^{3+}), anions (SO_4^{2-}, $S_2O_8^{2-}$ and BrO_3^-) and mixture of additives (e.g. $Fe^{2+}/Fe^{3+} + H_2O_2$). From thermodynamic point of view, most of these additives ought to be stronger electron acceptors than oxygen. These substances are added in order to scavenge conduction band electrons and generate oxidising radicals such as $^{\bullet}OH$. The redox and photochemical mechanisms for the generation of these radicals are well established by electrochemical studies. These mechanisms are relied upon for the explanation of rate enhancements in heterogeneous photocatalytic regimes. As electron scavenger, hydrogen peroxide may react as follows:

$$H_2O_2 + e^- \rightarrow {}^{\bullet}OH + {}^{-}OH \tag{5.21}$$

The anion $S_2O_8^{2-}$ can generate $SO_4^{\bullet -}$ by the absorption of light (Bao and Barker 1996). This photoreaction is shown by Eq. (5.22).

$$.S_2O_8{}^{2-} + h\nu \rightarrow SO_4{}^{2-} + SO_4{}^{\bullet -} \tag{5.22}$$

The $SO_4^{\bullet -}$ produced in Eq. (5.22) exhibits weak optical absorption in the UV-C region ($\lambda_{max} = 255$ nm) with $\epsilon_{max} = 1{,}000\ dm^3 mol^{-1} cm^{-1}$ (Neta and Huie 1985). The solvated $SO_4^{\bullet -}$ radical can readily generate hydroxyl radical as shown in Eq. (5.23).

$$SO_4{}^{\bullet -} + H_2O \rightarrow SO_4{}^{2-} + {}^{\bullet}OH + H^+ \tag{5.23}$$

The production of hydroxyl radical by the reactions above is responsible for the increase in photocatalytic degradation rate by the addition of anions. Bromate ion (BrO_3^-) is also known for its unrivalled potential for oxidation of organic compounds and the following mechanism has usually been suggested (Bahnemann et al. 2007):

$$BrO_3{}^- + 2H^+ + e_{CB}{}^- \rightarrow BrO_2{}^{\bullet} + H_2O \tag{5.24}$$

$$BrO_3{}^- + 6H^+ + 6e_{CB}{}^- \rightarrow [BrO_2{}^-, HOBr] \rightarrow Br^- + 3H_2O \tag{5.25}$$

Bromate can also decompose by the following pathway (Hong et al. 2002; Mills et al. 1996):

$$2BrO_3{}^- \rightarrow 2Br^- + 3O_2 \tag{5.26}$$

Excess BrO_3^- could diminish the rate of photocatalytic reaction owing to the adsorption of Br^- on the semiconductor photocatalyst (Sobana and Swaminathan 2007).

In ozonation-photocatalysis, trioxygen or ozone is dissolved in the heterogeneous photoreactivity solution. Ozonation-photocatalysis harvests the attractive features of ozone for organic pollutant removal which include the relatively high scavenging effect provided by ozone, ability to decrease the rate of de-excitation of electrons and elimination of residues (Hernández-Alonso et al. 2002; Jung and Park 2004). During the photocatalytic process, electron transfer to O_3 leads to increased generation of hydroxyl radical in reactivity systems. The chemical following reactions have been suggested to account for the enhancement of photocatalysis with ozone (Huang and Li 2011):

$$O_3 + e^- \rightarrow O_3{}^- \tag{5.27}$$

$$O_3{}^{\bullet} + H^+ \rightarrow HO_3{}^{\bullet} \tag{5.28}$$

$$HO_3{}^{\bullet} \rightarrow HO^{\bullet} + O_2 \tag{5.29}$$

Ozononation-photocatalysis provides double aptitude for the removal of recalcitrant organic pollutants. Application of such fortified methods could be utilised to overcome the failure of less rigorous processes such as ozonation-photolysis, as the latter was found to be less effective in the removal of N-nitrosodimethylamine (NDMA) from water (Mofidi et al. 2002).

Generally speaking, inorganic anions may have negative or positive effect on the rate of photocatalytic degradation depending upon the prevailing conditions. The orthophosphate HPO_4^{2-} has not been beneficial in any way to photocatalytic reactions. Doong et al. (2001) has observed inhibition of the photodegradation of monochlorophenol by HPO_4^{2-} on titanium dioxide surface. Similarly, in photodefluoridation of pentafluorobenzoic acid, the reaction rate was reduced by both HPO_4^{2-} and SO_4^{2-} and accordingly, the former demonstrated the highest inhibition (Ravichandran et al. 2007). The mechanism of rate inhibition by HPO_4^{2-} has not been clear but the blockage of the active sites of the semiconductor particles has recently been suggested (Gaya et al. 2009). The reduction in removal rate in presence of Cl^- and SO_4^{2-} has been attributed to the scavenging of hydroxyl radical by these anions as shown in Eqs. (5.54) and (5.55) (Djebbar and Sehili 1998).

$$SO_4{}^{2-} + {}^{\bullet}OH \rightarrow SO_4{}^{\bullet} + OH^- \tag{5.30}$$

$$Cl^- + {}^{\bullet}OH \rightarrow {}^-OH + Cl^{\bullet} \tag{5.31}$$

Even though $SO_4{}^{\bullet -}$ formed in the process is strongly oxidising, with reduction potential $E^o = 2.4$ (Bao and Barker 1996; Alipázaga et al. 2005), it is less reactive than hydroxyl radical which has $E^o = 2.7$ (Sawyer et al. 1995; Patnaik 2004). Similarly, H_2O_2 addition can improve photodestruction rate only at optimum doses. In excess H_2O_2, hydroxyl radicals in the system will be scavenged by excess peroxide as shown in the following equation (Chang et al. 2008):

$$H_2O_2 + {}^{\bullet}OH \rightarrow HO_2{}^{\bullet} + H_2O \tag{5.32}$$

As the resulting hydroperoxyl radical is less reactive than the hydroxyl radical consumed decline in photocatalytic degradation rate may be observed.

In fenton-photocatalysis, $Fe^{2+}/Fe^{3+} + H_2O_2$ are applied to assist the photocatalytic process. The addition of Fe^{3+} results in improved photocatalytic decomposition rate owing to the generation of hydroxyl radical. Two mechanisms have been suggested for the fenton mediation. The Fe^{3+} is believed that this cation is reduced to Fe^{2+} by conduction band electron on the surface of the semiconductor photocatalyst, a process which would reduce recombination of electron-hole pair, thus enhancing photocataltytic degradation. Secondly, an electron may be abstracted from hydroxyl ions to form the +2 oxidation state. These mechanisms are shown by Eq. (5.33) and (5.34).

$$Fe^{3+} + e_{CB}{}^{-} \rightarrow Fe^{2+} \tag{5.33}$$

$$Fe^{3+} + OH^{-} \rightarrow Fe^{2+} + OH^{\bullet} \tag{5.34}$$

Fe^{2+} can be formed by many other processes. In phenolic solutions, it has experimentally been shown that Fe^{2+} quickly forms as H_2O_2 is consumed, owing to the reduction of Fe^{3+} by the quinonic intermediates. This fact has been reaffirmed by the primary reactions in the fenton process as follows (Du et al. 2006; Yang et al. 2005b):

$$Fe^{2+} + H_2O_2 \rightarrow Fe^{3+} + OH^{-} + {}^{\bullet}OH \quad k = 51 - 58 dm^3 mol^{-1} s^{-1} \tag{5.35}$$

$$Fe^{3+} + H_2O_2 \rightarrow Fe^{2+} + H^{+} + {}^{\bullet}OOH \quad k = 0.001 - 0.02 dm^3 mol^{-1} s^{-1} \tag{5.36}$$

$$Fe^{2+} + H^{+} + {}^{\bullet}OOH \rightarrow Fe^{3+} + H_2O_2 \quad k = 1.2 \times 10^6 dm^3 mol^{-1} s^{-1} \tag{5.37}$$

$$Fe^{2+} + {}^{\bullet}OH \rightarrow Fe^{3+} + OH^{-} \quad k = 3.2 \times 10^8 dm^3 mol^{-1} s^{-1} \tag{5.38}$$

$$Fe^{2+} + 2H^{+} + O_2{}^{\bullet -} \rightarrow Fe^{3+} + H_2O_2 \quad k = 1.0 \times 10^7 dm^3 mol^{-1} s^{-1} \tag{5.39}$$

$$Fe^{3+} + O_2{}^{\bullet -} \rightarrow Fe^{2+} + O_2 \quad k = 5.0 \times 10^7 dm^3 mol^{-1} s^{-1} \tag{5.40}$$

5.3.2 *Photoelectrocatalysis*

Perhaps the most popular study of photoelectrodecomposition of pure substance was that reported by Fujishima and Honda some decades ago. These workers utilised a photoelectrolytic cell to split water into its component elements in their physical states. A schematic diagram of their pioneering cell is shown in Fig. 5.2. The Honda-Fujishima cell consisted of a clean rutile-TiO_2 photoanode and Pt counter electrode as cathode, each dipped into a half cell containing electrolyte. The half cells are connected by salt-bridge. By applying electric or chemical gradient oxygen is produced at the photoanode while hydrogen is produced at the cathode. The photoelectrochemical process may be envisaged as usual to start by light absorption, followed by the transfer of the photonic energy to a catalytically active site, where the dissociation reactions occur. The individual steps of this sequence comprise of absorption by the semiconductor photocatalyst, composite materials or dyes. The excitation energy transfer from the absorber to the catalyst is usually assumed to be an electronic conduction process, which, in less ordered semiconducting materials could also be due

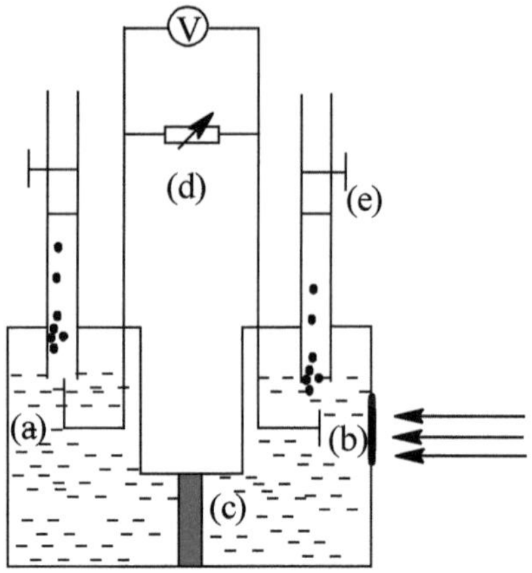

Fig. 5.2 Schematic diagram of Honda-Fujishima cell assembly. **a** Pt photoanode, **b** rutile cathode, **c** salt-bridge, **d** rheostat, **e** gas collector

to a hopping mechanism and dispersive transport of the light-induced excess electrons in the conduction band of semiconductor to its surface (Lewerenz et al. 2010). These photoprocesses together with electric energy produce synergistic effect in the degradation of chemical compounds.

The rate of photocatalytic degradation in combined operations depends on various parameters owing to the interplay of different processes. Apart from usual parameters that affect photocatalytic degradation such as concentration and pH, an important factor in photoelectrocatalysis is the applied potential. The Honda-Fujishima effect could not take place in absence of at least 0.2 V external bias (Fox 1983), it becomes clear that the the process may not be dubbed solely photocatalytic but photoelectrocatalytic. In fact the performance of photoelectrocatalytic process is normally higher than that of the conventional photocatalytic process. The photoelectrocatalytic degradation of methylene blue in aqueous solution was investigated by An et al. (2002) using three-dimensional electrode-photocatalytic reactor. The workers found that the methylene blue could be degraded more efficiently by photoelectrocatalytic process than by photocatalytic oxidation or electrochemical oxidation alone. The decolorization efficiency and chemical oxygen demand reduction were 95 % and 87 % for a photoelectrochemical process, respectively, while they were only 78 % and 68 % for a single electrochemical process and 89 % and 71 % for a single photochemical process.

There has been extensive development in the apparatus for use in photoelectrocatalysis. An electrode may consist of pure semiconductor photocatalyst or a suitable metal coated with the semiconductor photocatalyst. Cathode is usually made of inert material which range from the expensive Pt to the cheap graphite. While the photoanode is made of Ti metal usually coated with TiO_2 via dip-coating method. For example, Zainal et al. (2007) has studied the photoelectrocatalytic treatment of mixed dyes (consisting of six commercial dyes and textile effluents from cotton dyeing process) using TiO_2-coated Ti as anode. In TiO_2 photoelectrocatalysis, the TiO_2 must be in its crystalline phase as Özcan et al.'s recent study (2010) has reported no performance when TiO_2 obtained by thermal oxidation in air at 400–700 °C (confirmed to be amorphous TiO_2 by X-ray diffraction) was used in the absence of oxygen, bias or irradiation. Bao-xiu et al. (2007) studied the degradation of 2,4-dichlorophenol in a three-electrode photoelectrocatalytic oxidation reactor consisting of bielectrode anode (TiO_2/Ti and Fe rods) and graphite cathode. Some H_2O_2 was produced nearby the cathode and Fe^{2+} was continuously generated from Fe anode in solution when current and O_2 were supplied. The photoelectrolytic 2,4-dichlorophenol degradation process was synergistically assisted by electrofenton and peroxidisation. The degradation ratio of 2,4-dichlorophenol was 93 % in the integrative oxidation process, while it was only 31 % in elecrofenton process and 46 % in H_2O_2-assisted TiO_2 photoelectrocatalytic process.

5.3.3 *Sonophotocatalysis*

Ultrasound can facilely be applied in the enhancement of photocatalytic degradation. The mechanism by which ultrasound influences treatment processes is known as cavitation. Cavitation involves the nucleation, growth and implosive collapse of micro-bubbles or cavities occurring in fractions of microseconds and releasing large quantities of energy over a small location (Wang et al. 2011). The extreme temperature and pressure (up to 101.3 MPa and 4,727.0 °C has been attained respectively) can cause pyrolytic or combustion reactions (Peller et al. 2004), or induce the formation of oxidising species such as hydoxyl radical and other radical species through water scission (Bertelli and Selli 2004). The compression work occurs almost adiabatically—nearly no heat exchange with the surroundings takes place (Hiskia et al. 2001). We have earlier given a detailed account of sonophotocatalytic reactions in this chapter. The formation of radical species by ultrasonic irradiation can be represented by the following sonolysis equations:

$$H_2O))) \rightarrow H^{\bullet} + {}^{\bullet}OH \tag{5.41}$$

$$H^{\bullet} + O_2 \rightarrow HOO^{\bullet} \tag{5.42}$$

$$HOO^{\bullet} \rightarrow HO^{\bullet} + {}_{1/2}O_2 \tag{5.43}$$

$$O_2 \rightarrow 2O^{\bullet} \tag{5.44}$$

$$2HO^{\bullet} \rightarrow H_2O_2 \tag{5.45}$$

$$HOO^{\bullet} \rightarrow H_2O_2 + O_2 \tag{5.46}$$

Ordinary sonolysis may not be as effective as the sonophotocatalytic mineralisation of resistant pollutants even when combined with other advanced processes such as sonofenton or sonocatalysis (Ranjit et al. 2008). However, sonophotocatalysis has proven efficiency for destruction of many persistent pollutants such as dyes. In sonophotocatalysis, hydroxyl radical is generated from both ultrasonication and photocatalysis. In addition to these reactive radicals, positive hole is also produced which has the potential to directly cause the oxidation of organic substances. Both processes synchronise to yield enhanced photodegradation. Peller et al. (2003) have studied both sonolysis and sonophotocatalysis. These workers have shown that sonolysis is quite effective in the initial degradation of chlorinated aromatic molecules, but complete mineralization is difficult to achieve, while photocatalysis is selective toward the degradation of polar compounds but causes the build up of undesirable chemical intermediates. In contrast to sonolytic degradation, TiO_2 photocatalysis is very effective towards achieving complete mineralization almost by two fold. By simultaneously carrying out high frequency sonophotocatalysis these workers have succeeded in achieving faster and complete mineralization with no build up of toxic intermediates even at very low catalyst loadings. Both simultaneous and sequential combination of sonolysis and photocatalysis can have beneficial effect in enhancing the rate of degradation. Stock et al. (2000) have kinetically shown the additive character of the sonolytic and photocatalytic rate constant for the degradation of napthol blue black ($1.04 \times 10^{-2}\ min^{-1}$ and $0.56 \times 10^{-2}\ min^{-1}$, respectively). The rate constant for the combined method was $1.83 \times 10^{-2}\ min^{-1}$. On the whole, ultrasonication works perfectly well in providing synergism to photocatalysis on a small-scale. This is not denying the fact that challenge remains with respect to large-scale application due to the ineffective distribution of the cavitational activity over the reactor.

While discussing the concept of sonophotocatalysis it may be appropriate to bring the concept of sonopotential suggested by Marchante et al. (2009), which may find use in sonophotoelectrocatalysis. These workers defined sonopotential as the drop in the open circuit potential of a semiconductor nanostructured electrode caused by ultrasonic (US) irradiation. The sonopotential (V_{US}) is a function of ultrasonic power and may be expressed in analogous manner to photopotential as follows:

$$V_{US} = E - E_{US} \tag{5.47}$$

The sonopotential can further be expressed by Eq. (5.48).

$$V_{US} = m \frac{KT}{e} \ln \frac{n_{US}}{n} \tag{5.48}$$

where E is the open circuit potential and E_{US} is the open circuit potential upon ultrasonic irradiation, n and n_{US} are the concentrations of conduction band electrons in the absence and in the presence of ultrasonic irradiation, m is the electrode ideality factor, K the Boltzmann constant, T the temperature and e the elementary charge.

Several factors may affect the rate of sonophotocatalysis. One important variable is the ultrasonic frequency whose optimum value is required for effective sonophotocatalytic degradation. Recently, it has been shown that at low frequency (20 kHz) a lower liquid temperature favours the sonochemical degradation of 4-chlorophenol while at high frequency (500 kHz) the rate of 4-chlorophenol destruction is minimally perturbed with a slight optimum at about 40 °C. Above all, the maximum rate of degradation occurred only at 200 kHz (Jiang et al. 2006). Similarly, sonophotocatalysis can be affected by anions and ionic strength. Chen and Simirniotis (2002) have shown that Cl^- may assist the degradation of 4-chlorophenol, 2,4-dichlorophenol, and 2,4,6-trichlorophenol but excess of this anion can inhibit the action of sonophotocatalysis at low pH values. It is well known that in the degradation of chlorophenols the concentration of Cl^- in solution keeps increasing (Doong et al. 2001). Therefore, the inhibition of sonophotocatalysis in this case may be attributed to reduced solubility which may be as a result of common ion effect.

The possibility of improving sonochemical degradation by substituting air atmosphere with argon has been established. Nagata et al. (2000) have shown that the fraction of dilute aqueous solutions of 2-, 3- and 4-chlorophenol and pentachlorophenol degraded sonochemically in 1 h was higher in argon atmosphere (100 %) than in air (80–90 %). This may be attributed to the formation of more radical species in the sonochemical process in presence of argon. However, there was no study on the possibility of improving sonophotocatalytic degradation by inert atmosphere.

5.3.4 Microbially-aided Photocatalytic Processes

Photocatalysis may be used as destructive technology to improve traditional methods of treatment such as the biological methods. Generally, biological methods of treatment are inexpensive, operable in aerobic or anaerobic conditions but may be slow or result in partial mineralisation especially in the face of bio-refractory chemical compounds. The microorganisms used in biological treatment are usually pure cultures of aerobes, including *Pseudomonas sp.*, *Alcaligenes sp.*, *Azotobacter sp.*, *Rhodococcus sp.*, and *Cryptococcus sp*, but the inherent toxicity of non-biodegrable compounds or of the intermediates produced during the degradation process compromise the ability of such pure cultures to completely mineralise them (Goel et al. 2010). Even though biological treatment methods have a lot of advantages that are appealing to modern treatment technologies such as cost-effectiveness and yielding to hybrid strategies, they have the disadvantage of having to stand weeks for cultures to prepare.

Coupled photocatalysis-biological treatment methods of operation have the double aptitude of being efficient and capable of total removal of pollutant. Suryaman et al. (2006) considered coupling biological treatment with photocatalysis degradation method for the removal of phenol and to assess reduction in operational cost. These workers observed the shortening of mineralisation time to the biological treatment and the electrical cost was saved compared to the photocatalysis. Some workers considered pretreatment of bio-recalcitrant pollutants by photocatalysis prior to microbial treatment, in order to reduce the difficulty of mineralisation with the biological process (Yahiat et al. 2011b). For instance, Brossillon et al. (2008) have used photocatalysis as initial step in the degradation of azo dyes from Morroccan factory while biodegradation was used as a secondary step. The biodegradation of the azo dyes reactive black 5 and reactive yellow 145 confirmed the low biodegradability of these components under the influence of a strain of *Pseudomonas fluorescens*, cultivated at 25 °C, an initial pH of 7 and in presence of glucose as supplementary carbon source.

For an initial dye concentration of 40 $mgdm^{-3}$, the maximum yields of decolouration were 27 and 18 % for the reactive black and reactive yellow 145 respectively.

In another study by Yahiat et al. (2011a), the fungicide cyproconazole was also removed from water using *Pseudomonas fluorescens*, subsequent to photocatalytic destruction. The photocatalytic elimination of cyproconazole ($85 \times 10^{-3} gdm^{-3}$) was almost complete after 500 min of irradiation. In a span of 255 min, there was 85.8 % decontamination, 38.5 % mineralisation, 51.6 % oxidation and decrease in COD/TOC ratio. However, there was no decrease in toxicity as the EC50 values were 61 % initially and 64 % after 255 min irradiation. Clearly, this is an indication of the presence of by-products from photocatalysis after the biological treatment. Hence, it is recommended that the cheaper microbial treatment be let remove the major part of the pollutant and to have the more expensive photocatalysis to perform the final polish (L'Amour et al. 2008).

5.3.5 *Adsorption-photocatalysis*

Another innovative method for removal of treatment is the adsorption-photocatalysis. This integrated method can be applied especially in the removal of organic substances and inorganic anions. The method of operation can be simultaneous or sequential depending upon choice. In order to treat municipal wastewater as alternative secondary and tertiary treatment for wastewater reclamation, Vimonses et al. (2010) integrated sequentially a clay mixture fluidised-bed reactor and a TiO_2-impregnated kaolin annular slurry photoreactor as an adsorption-photocatalysis system. Primary effluent from sewage and secondary effluent from a membrane bioreactor treatment process were used to assess chemical removal capabilities of the clay reactor, the annular system and the hybrid process. The fluidised-bed reactor system demonstrated the prevailing removal efficiency toward PO_4^{3-} and NO_3^- and suspended solids. Complete elimination of PO_4^{3-} content was obtained in the adsorption stage, what could not have been achieved by photocatalysis considering record history of PO_4^{3-} as rate retardant. The TiO_2 annular reactor showed superior degradation of dissolved organic content. However, the presence of inorganic ions caused a detrimental effect on the performance of the latter owing to competitive adsorption on the photocatalyst. The integration of the adsorption and degradation system as a hybrid treatment process resulted in a synergetic enhancement for the chemical removal efficiency. Similarly, Cr^{6+} has been reduced by photoelectrochemical adsorption on red peanut skin as a post treatment technique and the adsorption was coupled to photocatalysis over $CuAl_2O_4/TiO_2$ hetero-system (Kebir et al. 2011). A reduction of more than 58 % of $HCrO_4$ was achieved under optimal conditions after about 2 h.

Many organic substances have been treated by the coupled adsorption-photocatalysis. Areerachakul et al. (2007) have reported a hybrid system of granular activated carbon fixed bed adsorption and the continuous TiO_2 photocatalysis system were sequentially used to treat metsulfuron-methyl, a herbicide, from waste water. Columns packed with granular activated carbon at different bed depths were operated at different filtration rates over a period of several weeks. Removal of metsulfuron-methyl via adsorption using the adsorption fixed beds of 5, 10 and 15 cm depths achieved a removal of 35, 55 and 65 % of metsulfuron-methyl respectively. In the continuous photocatalysis system, removal rates between 40 and 60 % were achieved. However, a high removal rate of over 90 % was realised with activated carbon-loaded TiO_2, lowering the retention time of the photocatalysis system to less than 10 min. Lastly, it has recently been suggested that the adsorption of inorganic anions can be more successful than their photocatalytic oxidation. Byrne and Mazyck (2009) applied sol-gel derived silica-titania composites to remove trace level mercury (100 $\mu g dm^{-3}$) from aqueous solutions by adsorption-photocatalysis and compared the removal efficiency with the precursors namely silica and Degussa P25 TiO_2. Under dark adsorption conditions approximately 90 % removal of mercury was realised with the silica-titania composites, which was comparable to that of Degussa P25. Silica that is devoid of TiO_2 performed poorly in comparison and was minimally affected by UV illumination. In actual fact, the UV irradiation appeared to cause the formation of Hg^0. The foregoing therefore underscores the combined adsorption-photocatalysis more than any of the single methods.

5.4 Energy-based Applications of Photocatalysis

5.4.1 Dye-sensitised Solar Cells

In 1939 photovoltaic effect was discovered by A. E. Becquerel (1788–1879), when a photocurrent was realised from an electrolytic cell consisting of platinum electrodes which were covered with AgCl or AgBr and immersed in illuminated aqueous solution. The low efficiency of the Bequerel's photovoltaic cell called for the use of new materials. In the 1950s to 1960s, solar cells developed rapidly owing to the use of crystalline Si (c-Si) and amorphous Si thin films in different energy scale applications such as space equipments. Currently, commercial c-Si solar cells for example have efficiency of up to 15 to 22 % (Razykov et al. 2011). In 1970s, photovoltaic cell research was refuelled by the energy crisis. Silicon has generally been the most commonly used material in solar cells. This material has been used in different forms such the amorphous form (a-Si), the monocrystalline form (mc-Si) and the polycrystalline (pc-Si) form.

Solar cells come in manufactured in many different forms and can generally be classified based on the composition of the photovoltaic devices as in Fig. 5.3. Silicon-based photovoltaic cells fall under the inorganic category. Other usually non-silicon based photovoltaic cells include those of binary atom type such as gallium arsenide (GaAs), the ternary type copper indium diselinide ($CuInSe_2$) and the quaternary type copper indium gallium selinide (CuInGaSe). Typical efficiencies of these inorganic photovoltaic cells range from 4-24 %. Several photovoltaic cells from inorganic-inorganic composites are available and include semiconductor-sensitised solar cells such as (CdSe/CdS sensitised TiO_2) (Poulose et al. 2012) and carbon based solar

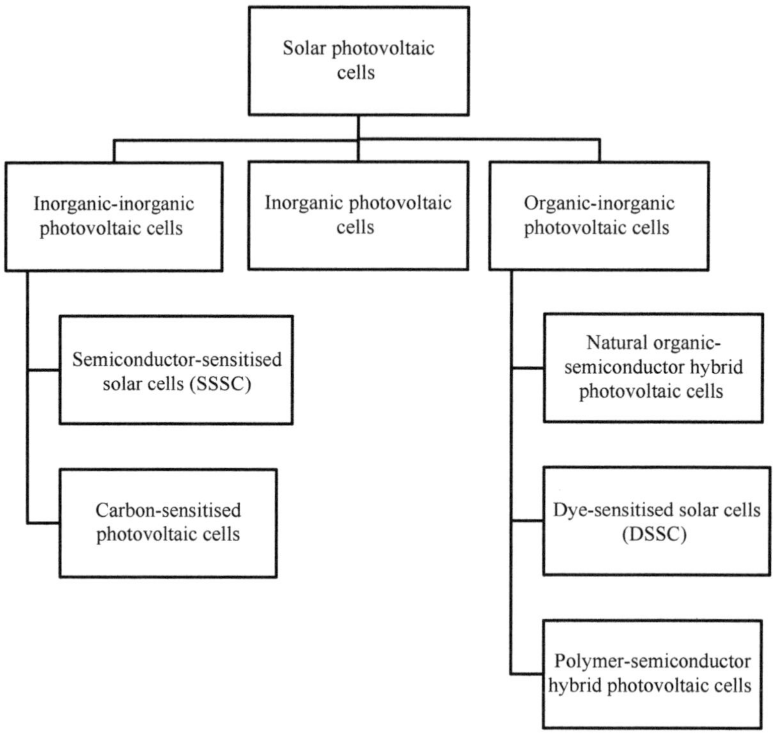

Fig. 5.3 Classification of solar photovoltaic cells

cells (such as C/Si and C/TiO_2) (Uddin et al. 2013). Organic-inorganic photovoltaics include those consisting of polymer blends (Hsu et al. 2009), natural organic pigments such as chlorophyll (De Padova et al. 2009; Amao and Yamada 2007) and the dye-sensitised solar cells (DSSC).

Owing to scarcity and cost of Si materials, the dye-sensitised solar cells have gained significant research interest. Basically, the light which irradiates the semiconductor-dye solution interface causes the excitation of the dye and subsequent transfer of electron to the conduction band of the semiconductor for electrical work. The electron injection into conduction band of the semiconductor photocatalyst has been confirmed by transient absorption measurement and laser photolysis experiments (Moser and Grätzel 1983; Huber et al. 2002). One of the requirements for a successful cell is a semiconductor with high degree of order, high stability against photocorrosion and suitable band gap energy for which mesoporous, nano-scale, *n*-type TiO_2 provides that end. Usually Pt is used as counter electrode. In Chap. 4 we have described the general principle of dye-sensitisation and mentioned that the dye is left in cationic form following electron injection. In DSSCs the neutral state of the dye is restored by a system of electrolyte of I^-/I_3^- redox couple. In systems of dye complexes, back transfer of an electron takes place from the semiconductor to the d-orbital of the metal (such as Zn^{2+}, Co^{2+} or Ru^{2+}). The solvents used for the electrolyte and the dye are either nitriles (e.g. acetonitrile and valeronitrile) or alkyl carbonates (such as ethylene carbonate and propylene carbonate used in lithium batteries).

$$I_3^- + 2e_{CB} \rightarrow 3I^- \tag{5.49}$$

Dye-sensitisation process can be likened to photosynthesis in the use of dye as solar absorber and sensitiser. An ideal sensitiser should be able to graft on the semiconductor surface, absorb below the threshold limit of 920 nm and inject electrons into the conduction band of the semiconductor with a quantum yield of 1 (Kong et al 2007). At this juncture we need to define the light harvesting efficiency of the dye φ_L which is given by:

$$\varphi_L = \left(1 - \frac{I}{I_o}\right) \tag{5.50}$$

where I_o and I is the intensity of the incident radiation and

Several sensitizers have been evaluated for use with TiO_2 in DSSC application which include fluoranthene-based organic dyes (with conversion efficiency of 1.33 %) (Wu et al. 2010), N, N-diphenylhydrazone-based metal-free organic dyes (efficiency = 7.74 %) (Shen et al. 2011) and complexes of Ru^{2+} with various derivatives of 2,2' bipyridine such as 2,2′,2″-terpyridyl −4,4′,4″-tricarboxylic acid and *cis*-di −(thiocyanato)bis(2,2′-bipyridyl)-4,4′-dicarboxylic acid. The sensitization of *n*-TiO_2 by Ru^{2+} complexes of has been studied in detail by Grätzel et al. in their Photonics and Interfaces Laboratory of the Ecole Polytechnique Federale de Lausanne (EPFL), Switzerland. Such studies using *cis*-Di −(thiocyanato)bis(2,2′-bipyridyl)-4,4′-dicarboxylate) ruthenium (II) (**14**) led to realisation of the most efficient cell known as the Grätzel cell which has efficiency of 11 % (Grätzel 2005).

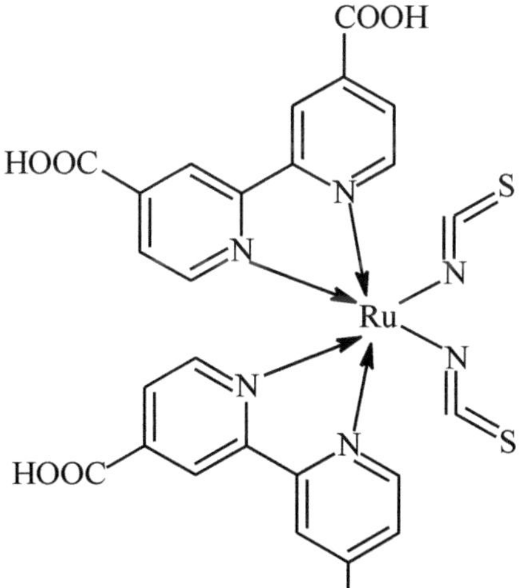

(**14**) (5.51)

Irradiation of the ruthenium complex causes excitation of the dye and electron injection into the conduction band of the TiO_2 which can be dispatched to the load connected to the system. The electron injection process involves a change in the oxidation state of the ruthenium as shown by Eq. (5.52).

$$cis - \text{Ru(II)(SCN)}_2(2,2' - \text{bipyridyl-4},4'\text{-dicarboxylate})_2 \rightarrow$$
$$cis - Ru\text{(III)(SCN)}_2(2,2'\text{-bipyridyl-4},4'\text{-dicarboxylate})_2 + \text{TiO}_2(\text{e}^-{}_{\text{CB}}) \quad (5.52)$$

In order to have greater stability, reduced electrolyte volume and compete with flexible solar cells such as Cu(In, Ga)Se_2 (CIGS) dye in the DSSCs or liquid junction solar cells has recently been substituted with quantum semiconductor sensitizers. Thus, solid-state DSSCs have been fabricated. In this case, the semiconductor sensitizer must be *p*-type so as to have a *p*/*n* heterojunction. Various *p*-type semiconductors such as CuI, CuSCN, CdS, PbS, PbSe, CdSe, InAs, CdTe, CdSeTe, Cu_2O have been utilised for this purpose. Similarly, quasi-solid state DSSCs have been fabricated by using electrolyte system of ionic liquids and the redox couple I^-/I_3^-. The thickness of the semiconductor l can be obtained from the laws of photometry if it is transparent to light at $\lambda > E_g$.

From Beer-Lamberts law the thickness l is given by:

$$l = \frac{1}{\varepsilon C} \log \frac{\text{I}_\text{o}}{\text{I}} \quad (5.53)$$

Equivalently, l can be derived from the light harvesting efficiency as shown by the following relationship:

$$\varphi_\text{L} = \left(1 - \frac{\text{I}_\text{o}}{\text{I}}\right) = (1 - 10^{-\varepsilon Cl}) \quad (5.54)$$

Where ε is the extinction coefficient of the adsorbed sensitizer, C is its concentration and l is the thickness of the absorbing.

Zinc oxide is also a good candidate for use in solar cells. Solar cells based on 2,2′-bipyridyl-4-4′-dicarboxylate sensitized ZnO gave 5 % efficiency as compared to, for example SnO_2 which has 1 % efficiency (Jayaweera et al. 2008). Even though Ru(II) 'bipyr' complexes are the best sensitizers for TiO_2 with an incident photon to current conversion efficiency (IPCE%, see Chap. 4) of 100 % between 510 and 570 nm, eosin Y showed the best performance for ZnO with an IPCE of 92 % (Wunderlisch et al. 2004). IPCE is proportional to the light harvesting efficiency of a photovoltaic cell. The equation relating these parameters is given as:

$$\text{IPCE} = \varphi_\text{L} \cdot \varphi_C \cdot \varphi_i \quad (5.55)$$

where φ_L is the light harvesting efficiency, φ_coll is the efficiency of electron collection while φ_i is the quantum yield for electron injection.

5.4.2 *Photocatalytic Hydrogen Production*

Photocatalytic is hydrogen production has been investigated primarily due to the concern on the depletion or unavailability of fossil fuels. Solar to H_2 fuel has

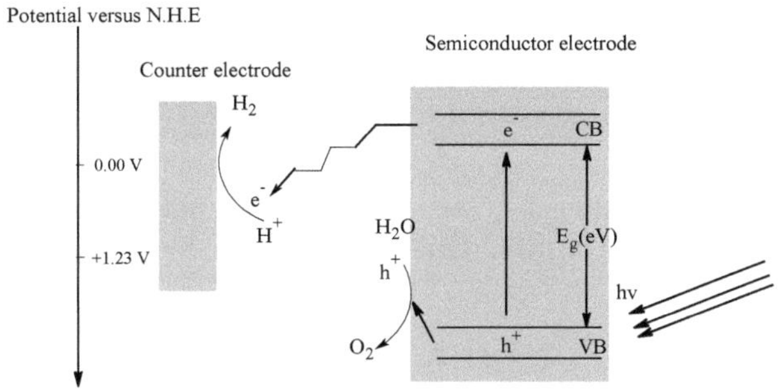

Fig. 5.4 Schematic diagram for solar hydrogen production

already attracted commitment of both the industrial community (such as the General Motors) and US Government (McCusker 2001). Semiconductor-assisted solar water splitting has been investigated as a source for hydrogen fuel. The use of semiconductor in this context was first introduced by Fujishima and Honda following came to limelight after the discovery of photosplitting of water described in Subsect. 5.3.2. The Fujishima-Honda cell consists of rutile TiO_2 anode and Pt counter electrode.

In photocatalytic water splitting, the semiconductor photocatalyst is irradiated by light having sufficient energy to cause charge separation whereupon the oxidation of water to oxygen takes place at the semiconductor interface and hydrogen ion is reduced to molecular hydrogen at the counter electrode. A schematic diagram showing the processes involved in photocatalytic hydrogen production is shown in Fig. 5.4. As a rule (see Chap. 1) for the reduction of the H^+ to take place the conduction band must be more negative (or lower) than the corresponding reduction potential of the H^+ whereas for the valence band must be more positive (or higher) than the oxidation potential of the H_2O adsorbed on the TiO_2. The water splitting reactions and corresponding potentials versus normal hydrogen electrode (N.H.E.) at pH 0 are shown by Eq. (5.56) and (5.57). The conduction band potential of TiO_2 is electrochemically suitable for the reduction of H^+ ions to H_2, since it lies above the potential required for the reduction reaction. However, the liberation of H_2 and O_2 is hindered by very high overpotentials (Swaddle 1990).

$$\text{At the counter electrode: } 2H^+ + 2e_{CB} \rightarrow H_2 \qquad E^o = 0.00\,V \tag{5.56}$$

$$\text{At the TiO}_2\text{ electrode: } 2H_2O + h^+ \rightarrow {}^1\!/_2O_2 + 2H^+ \qquad E^o = -1.23\,V \tag{5.57}$$

In addition, titanium dioxide is a wide-band gap semiconductor which absorbs mostly only the UV part of solar spectrum which corresponds to 5 %. An extension of the wavelength threshold of TiO_2 to the visible region can result in absorption of visible solar spectrum and conversion efficiency of up to 16 %. The maximum use of visible light would require a catalyst that absorbs below the band gap for water splitting (2.0 eV) which excludes even Bi_2WO_3 that has a band gap of 2.75 eV (Lai et al. 2013). Similarly, some visible light responsive semiconductor photocatalysts such as $BiVO_4$ and WO_3 cannot even split water. Cadmium sulphide satisfies the electronic requirements for good water splitting under visible light irradiation but this compound is electrochemically unsuitable as S^{2-} is preferentially oxidised by positive hole instead of H_2O (Abe 2010) as shown in Eq. (5.58).

$$CdS + 2h^+ \rightarrow Cd^{2+} + S \tag{5.58}$$

Thermodynamically, the water splitting reaction must be driven uphill, requiring a very large positive Gibbs free energy for the reaction to proceed as shown in Eq. (5.59).

$$2H_2O \rightarrow H_2 + {}_{1/2}O_2 \qquad \Delta G^o = +238 KJ\ mol^{-1} \tag{5.59}$$

The traditional photosplitting of water is dubbed "one-step water splitting" because a single photocatalyst is used. Undoubtedly, one-step photosplitting using TiO_2 is less efficient in hydrogen production despite the high photocatalytic activity of the semiconductor photocatalyst in environmental scenarios. Consequently, the search for efficient photocatalysts for water splitting and improvement of known photocatalysts was launched. Some of the strategies for improvement of the photocatalytic activity mimic those utilised for photocatalytic degradation. These strategies include doping, quantum confinement, and the preparation of oxide composites. For example, Fe ions can act as electron traps and this property extends to Fe-doped TiO_2 which can provide H_2 fuel at the rate of 15.5 μmol h^{-1} as opposed to 5.3 μmol h^{-1} with Cr-doped TiO_2 (Dholam et al. 2009). Improvement in the photocatalytic activity of CdS was obtained when quantum dots of the material were grown on graphitic carbon nitride nanosheets. The CdS (12 wt%)-C_3N_4 gave a specific hydrogen evolution rate of 4.494 mmol h^{-1} g^{-1} (Cao et al. 2013).

Some workers have achieved good results by the use of semiconductor photocatalyst and a sacrificial agent, which is usually an electron donor. For instance, *p*-Cu_2O with band gap 2.1 eV has produced considerable photocurrents in presence of the herbicide methyl viologen (Somasundaram et al. 2007). Similarly, certain composites of TiO_2 have also demonstrated significant activity for water splitting especially in the presence of sacrificial agents. In oxalic acid solution, $CuCr_2O_4/TiO_2$ composite at stoichiometric ratio of 10:7 has demonstrated improved performance over either the single TiO_2 or $CuCr_2O_4$ (Yan et al. 2009). Simamora et al. (2013) found that CuO (2.5 wt%)-TiO_2 composite can photosplit water and generate 3.1 μmol H_2 h^{-1} in presence of 10 mgdm^{-3} methylene blue within 5 h, but 99 % of the methylene blue degrades in the process. However, exceedingly higher amount of hydrogen (about 570 mmol h^{-1}) was produced by CuS/TiO_2 composite in presence of methanol, which corresponds to 32 times of the performance of P25-TiO_2 used under the same condition (Wang et al. 2013). Noble metal compounds and the metal loaded photocatalysts have also shown high photocatalytic activity for hydrogen production including Ag_2O(1 wt%)-loaded TiO_2 (H_2 production rate = 3,350 μmmol h^{-1}) (Lalitha et al. 2010), dye-sensitised or bare platinised TiO_2 (Kandiel et al. 2011) and gold-loaded TiO_2 (Rayalu et al. 2013).

Natural photosynthesis proceeds within the granum of plants by electron transfer mechanism known as the Z-scheme. As photocatalytic water splitting is designed for efficient solar energy utilisation, a revolutionary strategy for improving photocatalytic water splitting called the Z-scheme was recently introduced. The

Z-scheme is a two-step water splitting system consisting of two photocatalysts—one for H_2 liberation and the other for O_2 evolution—along with IO_3^-/I^- as a redox shuttle as mediator. Currently, several photocatalyst systems are being designed for use as Z-scheme type materials towards effective evolution of H_2 and O_2 (Abe 2010) Ding et al. 2013;. The two-step excitation process in Z-scheme allows the use of semiconductor photocatalysts that do not satisfy the band gap requirement for water splitting. For example, Pt/SrTiO3:Rh can be utilised as H_2-photocatalyst while $BiVO_4$ or WO_3 is charged with O_2 evolution. Typical Z-scheme type photocatalysts have been furnished by Sasaki et al. (2008) in which either Pt, Rh or Ru was used as co-catalyst. For H_2 production $SrTiO_3$:Rh was used while for oxygen evolution $BiVO_4$ was used photocatalyst. Meanwhile, Fe^{3+}/Fe^{2+} was used as mediator. It was found that the activity for water splitting did not vary with co-catalyst. However, the (Ru/$SrTiO_3$:Rh)-($BiVO_4$)-(Fe^{3+}/Fe^{2+}) photocatalysis system gave a quantum yield of 0.3 % and a stable activity for more than 70 h.

5.5 Recent Topics in Photocatalysis

5.5.1 Photocatalytic Oxidation of Inorganic Cations

Research thrust in photocatalytic degradation has been mostly geared towards the depollution of organic substances. Relatively little is explored about the photocatalytic oxidation of inorganic anions and its mechanism. However, the prospects of semiconductor photocatalysts in the abatement of inorganic pollutants of various forms including acidic oxides is well established (Faust et al. 1989). Over the past few years, attention has been given to the abatement of aquatic arsenic by different advanced oxidation methods owing to its unrivalled toxicity (Hug and Leupin 2003). Arsenic may be derived from organic arsenic sources owing to uncontrolled use of arsenic compounds. For example, phenyl substituted arsenic compounds are widely used as feed additives in the poultry industry and these compounds are currently becoming of serious environmental concern (Zheng et al. 2010). Inorganic arsenic species are more toxic than organic ones, among which As^{+3} is in the forefront (Sharma and Sohn 2009). Basically, the element arsenic exists in four oxidation states: the $+5$ oxidation state or arsenate, the $+3$ known as the arsenite, the 0 oxidation state popularly known as the arsenic and the -3 oxidation state referred to as arsine. The arsenite and the arsenate are common in natural waters—the arsenite (As^{3+}) include $As(OH)_3$, $As(OH)_4^-$, AsO_2OH^{2-} and AsO_3^{3-} while the arsenate species (As^{5+}) being in the form of AsO_4^{3-}, $HAsO_4^{2-}$, $H_2AsO_4^-$ (Mohan and Pittman 2007).

The photocatalytic oxidation of the hazardous As^{+3} has recently been demonstrated (Yang et al. 1999; Ferguson and Hering 2006; Li et al. 2009; Pena et al. 2005; Bissen et al. 2001). When arsenite is oxidised to arsenate, the toxicity of the original trivalent cation is drastically reduced. Even though the photocatalytic oxidation of arsenite is established, the oxidation kinetics of As^{+3} and mechanism with both natural and technical oxidants has not been clearly unveiled (Hug and Leupin 2003). Lee and Choi (2002a) attempted for the first time a detailed investigation of the mechanism of As^{+3} photocatalytic oxidation to As^{+5} using irradiated TiO_2 suspension. Their study in both UV and visible light stressed that superoxides are responsible for the oxidation of As^{+3} (Lee and Choi 2002a; Ryu and Choi 2004). In a photoelectrochemical study, these workers also attributed the observed decrease in photoanodic current at TiO_2 electrode to the oxidation of As^{+3} by superoxides (Ryu and Choi 2006).

$$As^{+3} + O_2{}^{-\bullet}/\,HO^{\bullet}/\,h^{+}(?) \rightarrow As^{+5} \tag{5.60}$$

Currently, As^{+3} photocatalytic oxidation has attracted special interest particularly as its mechanism is faced with the controversy of which oxidant is dominant (Eq. (5.60)) (Leng et al. 2007; Ryu and Choi 2007). Several studies have been carried out to confirm the major photo-oxidant in As^{+3} oxidation. The majority of the studies disputed the significance of the superoxide mechanism. Yoon and Lee (2005) argued that positive hole is responsible for As^{+3} oxidation based on the fact that excess methanol, a hydroxyl radical scavenger, did not retard As^{+3} oxidation rate while the addition of oxalate, a hole scavenger, suppressed the photocatalytic oxidation rate. Dutta et al. (2005) investigated the photocatalytic oxidation of As^{+3} with benzoic acid in solution and found in support of hydroxyl radical mechanism, that the rate constant of the oxidation of As^{+3} by $O_2{}^{-\bullet}$ is three orders of magnitude slower than the rate constant of benzoic acid reaction with $HO^{\bullet}$. In this study, the photocatalytic oxidation of As^{+3} took place in minutes and followed zero order kinetics.

Perhaps one vital question that has not been dealt with by the previous workers was the direct rate parameters concerning both As^{+3} and superoxide. Xu et al. (2005) determined the rate constant for the reaction of As^{+3} with $O_2{}^{-\bullet}$ (3.6×10^6 moldm^{-3} s^{-1}) over TiO_2 by pulse radiolysis and competition kinetics. Based on transient absorption decay as a function of As^{+3} concentration these workers concluded that adsorbed hydroxyl radical/trapped holes play a central role in the photocatalytic oxidation. Most recently, Yoon et al. (2009) have provided a conclusive evidence to support the opinion that hydroxyl radical and valence band hole are the main oxidants in the photocatalytic oxidation of As^{+3} over TiO_2. These workers investigated the correlation between As^{+3} oxidation and superoxide by measuring both As^{+5} and H_2O_2 simultaneously. Their finding indicated inhibition of As^{+3} oxidation with increase in H_2O_2 generation which confirms the insignificance of $O_2{}^{-\bullet}/HO_2{}^{\bullet}$ oxidation. The progression in research therefore favours the fact that hydroxyl radical/holes are responsible for As^{3+} oxidation. It may be logical therefore to conclude that all oxidation pathways are driven by these reactive oxidants.

5.5.2 *Remote Photocatalytic-oxidation*

Even though oxidising species generated on semiconductor photocatalyst persist in adsorbed phase, they may desorb, diffuse and react in a fluid phase (Turchi and Ollis 1990; Kubo et al. 2004; Vincent et al. 2008). Evidence for the diffusion of $HO^{\bullet}$ into gas phase has been established (Murakami et al. 2006) based on the quantum

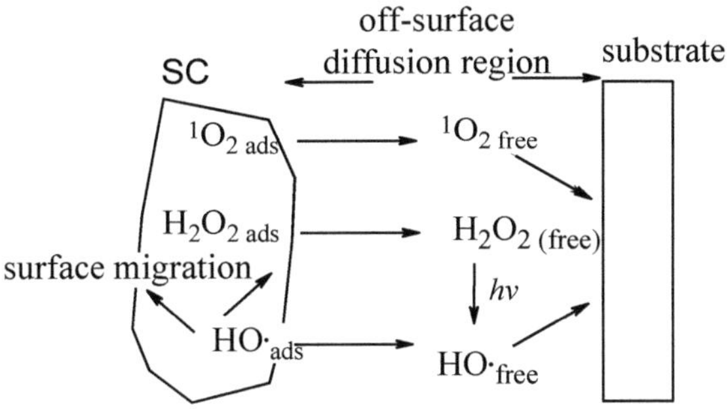

Fig. 5.5 A schematic diagram showing the diffusion of active species on and off semiconductor surface, and remote oxidation

yield (5×10^{-5}) for the generation of the radical. However, the mechanism of $HO^{\bullet}$ desorption is not very clear. Murakami (2007) suggested a mechanism based on the observation of titania surface-based peroxo-species as intermediates of oxygen photoevolution on the surface of TiO_2 (Eqs. (5.61), (5.62), (5.63), (5.64) and (5.65)). Another plausible mechanism involves the reduction of H_2O_2 as shown by Eq. (5.65) (Sun and Bolton 1996; Nakamura and Nakato 2004).

$$[Ti - O - Ti]_s + h^+ + H_2O \rightarrow [Ti - O \bullet HO - Ti]_s + H^+ \tag{5.61}$$

$$[Ti - O \cdot HO - Ti]_s \rightarrow [Ti - O - O - Ti]_s + H^+ \tag{5.62}$$

$$[Ti - O - O - Ti]_s + H_2O \rightarrow [Ti - O - O - H \bullet HO - Ti]_s \tag{5.63}$$

$$[Ti - O - O - H \bullet HO - Ti]_s \rightarrow [TiO \bullet HO - Ti]_s + HO^{\bullet} \tag{5.64}$$

$$H_2O_{2ads} + e^-{}_{CB} \rightarrow {}^{\bullet}OH + OH^- \tag{5.65}$$

In recent years, the study of the mobility of reactive oxygen species in photocatalytic oxidation systems has attracted much interest. Tatsuma et al. (1999) demonstrated for the first time the remote photocatalytic activity of TiO_2 for eradication of pollutant. These workers confirmed the attack of oxidising species of a remote methylene blue coated glass situated 500 μm away from a TiO_2-coated glass in an aerated environment. It appears more research into this area may breed amazing applications. Currently, a novel application of remote oxidation for solid surface photo-patterning has been developed (Tatsuma et al. 2002; Kubo et al. 2004). A TiO_2-coated photomask is placed on the material to be patterned and irradiated by a light source within 100 μm away.

The mobility reactive species is not restricted to fluid phase only as surface migration of reactive species as far as 20 μm has been demonstrated by Haick and Paz (2001) by measuring the degradation of different types of self-assembled monolayers anchored near TiO_2. Lee and Choi (2002b) also observed the decomposition of immobile hexane soot layer on TiO_2 film upon UV irradiation. Although a negligible portion (2 μm thick) was mineralised in 30 h the study could lead to potential non-environmental applications of photocatalysis. Figure 5.5 shows the diffusion mechanism involved in remote oxidation.

It is not clear whether all reactive oxygen species participate in remote oxidation mechanism as they do in the conventional photocatalytic process. However, it is known that ionic radical adsorbates on metal oxide semiconductors have restricted mobility (Ishibashi et al. 1998; Ishibashi et al. 2000). Recently however, Kwon and Yoon (2009) asserted that $O_2^{-\bullet}/HO_2^{\bullet}$ are mobile. The H_2O_2 generated on the surface of TiO_2 has also demonstrated the ability to diffuse to gas phase with quantum yield standing at 1.4×10^{-7} (Kubo and Tatsuma 2004). When the TiO_2 surface is saturated H_2O_2 cannot decompose but probably diffuse into the immediate medium. The amount of H_2O_2 released to the gas phase diminished as the thickness of the TiO_2 coating is reduced and no H_2O_2 was observed when dry air or N_2 was used. Diffused singlet oxygen molecule from TiO_2 surface has also been detected by total internal reflection fluorescence microscopy (Naito et al. 2006). However, singlet oxygen oxidation is not significant even in the conventional photocatalytic oxidation (Zhang et al. 1998).

One vital question which would have to be answered is how to improve the remote activity of TiO_2 since the process is relatively much slower than conventional photocatalytic oxidation (Tatsuma and Kubo 2007). Kubo and Tatsuma (2005) showed that Pt-loaded TiO_2 exhibited the highest activity among anatase TiO_2, oxygen deficient TiO_2, brookite TiO_2, ZnO, WO_3, metal (Au, Pt, Ag and Pd)-loaded anatase TiO_2 and metal (Au, Pt and Ag)-loaded ZnO. Their study also revealed that the conventional photocatalytic oxidation is a measure of remote photocatalytic oxidation. Other workers (Park and Choi 2004) also demonstrated the possibility of enhancing the mobility of $HO^{\bullet}$ by surface fluorination since the existence of $HO^{\bullet}_{ads}/h^{+}$ would be allowed as shown by Eq. (5.66). In fact the mechanism of phenol degradation on fluorinated titania has been based on reaction with free $^{\bullet}OH$ in solution (Minero et al. 2000a; Minero et al. 2000b).

$$\equiv Ti - F + OH^{-} + h^{+} \rightarrow \equiv TiF + {}^{\bullet}OH_{free} + H^{+} \tag{5.66}$$

5.5.3 Dual-cycle Photocatalysis

Certain industrially important processes release both organic and inorganic by-products to the enviironment. The treatment of both contaminants may become imperative. Following an investigation of the efficacy of TiO_2 photocatalysis for treating water contaminated with several dissolved metals specifically Ag, Au, Cd, Cr, Cu, Hg, Ni, Pt, and a variety of organic compounds which include ethylenediaminetetraacetate, methanol, phenol and nitrobenzene, Prairie et al. (1993) recommended that both oxidation and reduction processes be taken into account in the design of efficient photocatalytic system. Such study will be vitally important since field samples such as tannery and textile effluents contain collection of both organic and inorganic compounds mixed in solution. However, there have been only a few reports so far depicting the simultaneous abatement of organic and inorganic pollutants by photocatalytic redox process.

Table 5.1 Experimental observations on the simultaneous and synergistic conversion of some dyes and metal ions. (Deduced from reference Mills et al. 2003)

Test	TiO_2	Ag^+	Cr^{6+}	Rhodamine B	Acid orange 7	Inference
1	+	+	−	−	−	Ag^+ was not reduced in absence of the dye
2	+	+	−	+	−	Both dye and metal ion were converted
3	−	+	−	+	−	Both dye and Ag^+ was deposited on TiO_2 as Ag (size < 0.1 μm)
4	−	−	−	−	+	In these cases, both dye and Cr^{6+} were converted. The conversion of the latter in the absence of TiO_2 may imply the formation of complex
5	−	+	−	−	+	
6	−	+	−	−	+	
7	−	−	+	−	+	Insignificant conversion of dye and metal ion
8	+	−	−	−	+	Accumulation of H_2O_2. 30 μm of H_2O_2 was produced after 4 h
9	+	−	+	−	+	H_2O_2 production was insignificant
10	−	−	+	+	−	Negligible conversion of Cr^{6+} or complexation

− and + used in the table show absence and presence of component respectively

The efficacy of simultaneous decontamination of ethyl lactate and phenol together with copper ions by UV/TiO_2 photocatalytic degradation was investigated by Zou et al. (2007). These pollutants were chosen by the authors as model owing to their widespread presence in wastewater discharged by semiconductor manufacturing facilities. At optimum conditions (which corresponded to 0.1 gdm^{-3} TiO_2, pH 3–4, [phenol] = [ethyl acetate] = 1×10^{-3} dm^3), the heterogeneous photocatalysis is able to simultaneously remove both copper and organic compounds. Green solids were observed in solution and this was assigned to $CuCO_3$. The authors observed higher organic decontamination with higher concentration of Cu^{2+}. This is an indication of the positive influence of Cu^{2+} possibly via cupric-based fenton reaction. In anoxic conditions, copper removal was inhibited which re-emphasizes the significance of the presence of oxygen in photocatalytic reactions.

Sun et al. (2005) found that the presence of Cr^{6+} in TiO_2 suspensions irradiated with visible light can promote the removal of the 4-chlorophenol and the amount removed was proportional to the initial concentration of the Cr^{6+}. There was negligible or minimal interaction of 4-chlorophenol with the Cr^{6+} but it was observed that an optimal concentrations of Cr^{6+} and TiO_2 exist for the Cr^{6+} reduction and 4-chlorophenol oxidation. Cations such as Cu^{2+}, Fe^{3+}, Mn^{4+}, Ce^{4+} and V^{5+} did not promote the photocatalytic oxidation of 4-chlorophenol. Strikingly similar synergistic effects were reported by Kyung et al. (2005) in the simultaneous conversion of dyes and heavy metal ions in aqueous ternary component systems consisting of TiO_2/dye/metal ion under visible light ($\lambda > 420$ nm). TiO_2/Cr^{6+}/acid orange 7, TiO_2/Cr^{6+}/rhodamine B, TiO_2/Ag^+/acid orange 7, and TiO_2/Ag^+/rhodamine B were chosen as test systems. Table 5.1 summarises the study to investigate the abatement of acid orange 7 or rhodamine B in conjuction with Ag^+ or Cr^{6+}.

The results in Table 5.1 revealed marked increase in the removal rates of dyes in the presence of metal ions. Unlike in the foregoing cases, the reduction rates of metal ions in visible-light-illuminated TiO_2 suspensions were negligible, but they were highly accelerated with dyes present especially in TiO_2/Cr^{6+}/acid orange 7 system which is an indication of synergism.

5.5.4 *Intelligent Colorimetric Probe for Functionalised TiO_2 Surfaces*

Since the introduction of the first self-cleaning glass (Activ^{TM}) by Pilkington in 2001, a wide variety of commercial TiO_2-modified glasses have appeared in the market with various trademark names as Sunclean^{TM} (from PPG), Neat^{TM} (CGI) and Bioclean^{TM} (St. Gobain), and numerous functionalised tiles have been manufactured (Mills et al. 2005). The self-cleaning of these glasses is based on the double surface action of TiO_2 coatings (5-20 nm) to degrade organic pollutants by photocatalytic processes and to allow easier washing of surface by inducing photo-hydrophilicity. An outstanding problem had been the development of commercially effective test method for the photoactivity of self-cleaning titania films. Earlier methods were based on the mineralisation of stearic acid Eq. (5.67) and monitoring disappearance or the formation of mineralisation products using chromatography, ellipsometry, fourier-transform infrared spectroscopy or by the use of indicator solution. The latter were kinetically very slow and therefore unsuitable for industrial purposes (Mills et al. 2008; Augugliaro et al. 2010).

$$CH_3(CH_2)_{16}COOH \rightarrow 18CO_2 + 18H_2O \tag{5.67}$$

Mills et al. (2005) provided and named "intelligent ink" with good reason, an ideal, inexpensive formulation to evaluate rapidly, at different scales, the photocatalytic activity of TiO_2 thin films. A typical intelligent indicator contains a polymer, usually aqueous solution of hydroxyethyl cellulose, glycerol and a redox dye. The pioneering dye is resazurin (blue). When resazurin is printed on Pilkington's Activ self-cleaning films and then illuminated it gets reduced irreversibly to resorufin (pink) and further reduced form, the dihydroresorufin which can be oxidised back to resorufin in presence of oxygen and other colourless products of mineralisation (Mills et al. 2006). This action of photocatalytic bleaching can be readily noticed in a few minutes. Mills and McGrady (2008) therefore compared the performance of resazurin with resorufin, 2,6-dichloroindophenol and methylene blue for the rapid assessment of the activity of very thin, photocatalyst films, such as Activ^{TM} self-cleaning glass. The order of photocatalytic bleaching of ink indicator was as follows:
methylene blue > resazurin > 2,6-dichloroindophenol > resorufin

Despite the relatively rapid rate of photobleaching for methylene blue its reduced form is oxygen-sensitive that the ink cannot be photoreduced under aerobic conditions, thus rendering the ink unsuitable for industrial application. Similarly, the photoreduction kinetics of resazurin is affected by the amount of oxygen present and by relative humidity. The reduction of resorufin was the slowest. Hence, these observations suggest choice of 2,6-dichloroindophenol for the assessment of the photoactivity of TiO_2 thin films. Eq. (5.68) shows the conversion of 2,6-dichloroindophenol or (4-{[(1Z)-3-chloro-5-methyl-4-oxocyclohexa-2,5-dien-1-ylidene]amino}phenolate, into its conjugate base the sodium 4-[(3-chloro-4-hydroxy-5-methylcyclohexa-2,5-dien-1-yl)amino]phenolate upon photobleaching over TiO_2 thin films. Even though photoreduction of methylene blue may not occur in aerobic condition this compound remains a useful indicator in this context especially at laboratory scale.

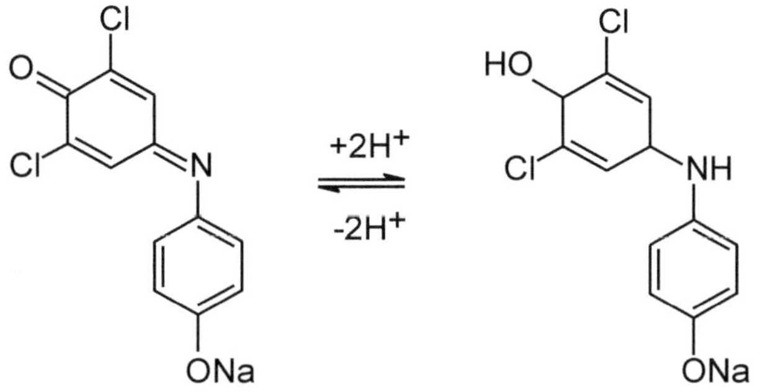

(5.68)

Instead of exploiting photocatalytic reduction, the oxidative degradation of methylene blue may suffice to assess the activity of very thin (crystallite sizes of 5–10 nm) TiO_2 films on self-cleaning glasses even with low intensity light sources (1 $mWcm^{-2}$) (Fateh et al. 2011). Chin and Ollis (2007) had investigated the kinetics of photodecolourisation of acid blue 9 and reactive black 5 on Pilkington Activ™ glass comprising 18 nm TiO_2 crystallites, using UV-A irradiance of 1.4 $mWcm^{-2}$. These workers described a novel horizontal coating method for thin layer dye deposition that minimizes the amount of solution used while forming a fairly uniform dye layer. The kinetics of photodecolorisation of acid blue 9 and reactive black 5 agreed with a two-step series reaction model with rate constants in the orders of 10^{-2} and 10^{-3}, respectively, signifying the irreversible formation of coloured intermediate and the final bleached product. Each reaction step follows a simple irreversible first-order reaction rate form. Although the use of these dyes may require hours, their applicability for the assessment of photoactivity of self-cleaning glasses has proven to be successful.

5.6 Conclusion

Crucial information about the experimental approaches such as reaction conditions and reactors used in photocatalytic degradation may not be readily and comprehensively apparent to many researchers. Regular highlight of these information are required to bridge the missing links. This chapter has not exhaustively provided insights into the approaches at disposal for the photocatalytic decontamination of environmentally important pollutants, but may be of some benefit to many readers.

In order to improve efficiency of operation, ordinary semiconductor photocatalysis may be combined either with physical operation or chemical system. This practice usually results in efficiency that will not have been realised using any of the methods singly. One minor area that is still appealing for further research is the assessment of specific toxicity of pollutants under semiconductor photocatalysis, despite the fact that the major photocatalysts, specifically TiO_2 and ZnO, have been used as tooth fillers thanks to their benign nature. Arguably, biological treatment may lessen the amount of toxic products that may be formed if photocatalytic treatment is operated singly.

Global need for alternative energy has increased the quest to develop photocatalytic sources which include photocatalytic hydrogen production and electricity generation. The dye-sensitisation method of producing electricity appears to be comparatively more successful despite the old age of photocatalytic hydrogen production since there are more of the products based on the former than those based on the latter. It is therefore hoped that future approaches such as the use of Z-scheme type photocatalysts will contribute to the realisation of relatively efficient hydrogen production systems.

References

Abdullah MDP, Nainggolan H (1991) Phenolic water pollutants in a Malaysian river Basin. Env Monit Assess 19:423–431

Abe R (2010) Recent progress on photocatalytic and photoelectrochemical water splitting under visible light irradiation. J Photochem Photobiol C 11:179–209

Abramović BF, Anderluh VB, Topalov AS (2004) Kinetics of photocatalytic removal of 2-amino-5-chloropyridine from water. APTEFF 35:1–80

Ali MF, El-Ali BM, Speight JG (2005) Handbook of industrial chemistry. McGraw-Hill, United States of America

Alipázaga MV, Moreno RGM, Linares E et al (2005) Oxidative DNA damage induced by autoxidation of microquantities of S(IV) in the presence of Ni(II)-Gly-Gly-His. Dalton Trans 44:3738–3744

Amao Y, Yamada Y (2007) Photovoltaic conversion using Zn chlorophyll derivative assembled in hydrophobic domain onto nanocrystalline TiO_2 electrode. Biosens Bioelectron 22:1561–1565

An T, Zhang M, Wang X et al (2005) Photocatalytic degradation of gaseous trichloroethene using immobilized ZnO/SnO_2 coupled oxide in a flow-through photocatalytic reactor. J Chem Technol Biotechnol 80:251–258

An T-C, Zhu X-H, Xiong Y (2002) Feasibility study of photoelectrochemical degradation of methylene blue with three-dimensional electrode-photocatalytic reactor. Chemosphere 46:897–903

Araa J, Nieto JLM, Melián JAH et al (2004) Photocatalytic degradation of formaldehyde containing wastewater from veterinarian laboratories. Chemosphere 55:893–904

Areerachakul N, Vigneswaran S, Ngo HH et al (2007) Granular activated carbon (GAC) adsorption-photocatalysis hybrid system in the removal of herbicide from water. Sep Purif Technol 55:206–211

Augugliaro V, Coluccia S, Loddo V et al (1999) Photocatalytic oxidation of gaseous toluene on anatase TiO_2 catalyst: mechanistic aspects and FT-IR investigation. Appl Catal B 20:15–27

Augugliaro V, Litter M, Palmisano L et al (2006) The combination of heterogeneous photocatalysis with chemical and physical operations: A tool for improving the photoprocess performance. J Photochem Photobiol A 7:127–144

Augugliaro V, Loddo V, Palmisano G et al (2010) Clean by light irradiation practical applications of supported TiO_2. Royal Society of Chemistry, United Kingdom
Bahnemann W, Muneer M, Haque MM (2007) Titanium dioxide-mediated photocatalysed degradation of few selected organic pollutants in aqueous suspensions. Catal Today 124:133–148
Bao Z-C, Barker JR (1996) Temperature and ionic strength effects on some reactions involving sulfate radical [SO_4^-(aq)]. J Phys Chem 100:9780–9787
Bao-xiu Z, Xiang-zhong L, Peng W (2007) Degradation of 2,4-dichlorophenol with a novel TiO_2/Ti-Fe-graphite felt photoelectrocatalytic oxidation process. J Environ Sci 19:1020–1024
Barceló D, Hennion MC (2008) Trace determination of pesticidesand their degradation products in water, techniques and instrumentation in analytical chemistry. Elsevier, Amsterdam
Bertelli M, Selli E (2004) Kinetic analysis on the combined use of photocatalysis, H_2O_2 photolysis, and sonolysis in the degradation of methyl tert-butyl ether. Appl Catal B 52:205–212
Bianchi CL, Pirola C, Ragaini V (2006) Mechanism and efficiency of atrazine degradation under combined oxidation processes. Appl Catal B 64:131–138
Bissen M, Vieillard-Baron M-M, Schindelin AJ et al (2001) TiO_2-catalyzed photooxidation of arsenite to arsenate in aqueous samples. Chemosphere 44:751–757
Bláha L, Klánová J Klán P et al (2004) Toxicity increases in ice containing monochlorophenols upon photolysis: environmental consequences. Environ Sci Technol 38:2873–2878
Blount MC, Falconer JL (2002) Steady-state surface species during toluene photocatalysis. Appl Catal B 39:39–50
Boulamanti AK, Philippopoulos CJ (2009) Photocatalytic degradation of C_5–C_7 alkanes in the gas–phase. Atmos Environ 43:3168–3174
Brillas E, Calpe JC, Casado J (2000) Mineralization of 2,4-D by advanced electrochemical oxidation processes. Water Res 34:2253–2262
Brosillon S, Djelal H, Merienne N et al (2008) Innovative integrated process for the treatment of azo dyes: coupling of photocatalysis and biological treatment. Desalination 222:331–339
Buechler KJ, Noble RD, Koval CA et al (1999) Investigation of the effects of controlled periodic illumination on the oxidation of gaseous trichloroethylene using a thin film of TiO_2. Ind Eng Chem Res 38:892–896
Byrne HE, Mazyck DW (2009) Removal of trace level aqueous mercury by adsorption and photocatalysis on silica-titania composites. J Hazard Mater 170:915–919
Calza P, Massolino C, Monaco G et al (2008) Study of the photolytic and photocatalytic transformation of amiloride in water. J Pharm Biomed Anal 48:315–320
Cao S-W, Yuan Y-P, Fang J et al (2013) In-situ growth of CdS quantum dots on g-C_3N_4 nanosheets for highly efficient photocatalytic hydrogen generation under visible light irradiation. Int J Hydrogen Energy 38:1258–1266
Castiglioni S, Fanelli R, Pomati F et al (2006) Removal of pharmaceuticals in sewage treatment plants in Italy. Environ Sci Technol 40:357–363
Chang M-W, Chen T-S, Chern J-M (2008) Initial degradation rate of p-nitrophenol in aqueous solution by fenton reaction. Ind Eng Chem Res 47:8533–8541
Chatzitakis A, Berberidou C, Paspaltsis I et al (2008) Photocatalytic degradation and drug activity reduction of Chloramphenicol. Water Res 42:386–394
Chen Y-C, Smirniotis P (2002) Enhancement of photocatalytic degradation of phenol and chlorophenols by ultrasound. Ind Eng Chem Res 41:5958–5965
Chen LC, Huang C-M, Hsiao M-C et al (2010) Mixture design optimization of the composition of S, C, SnO_2-codoped TiO_2 for degradation of phenol under visible light. Chem Eng J 165:482–489
Chin P, Ollis DF (2007) Decolorization of organic dyes on Pilkington Activ^{TM} photocatalytic glass. Catal Today 123:177–188
Chiron S, Minero C, Evione D (2007) Occurrence of 2,4-dichlorophenol and of 2,4-dichloro-6-nitrophenol in the Rhône River Delta (Southern France). Environ Sci Technol 41:3127–3133
Choi W, Ko JY, Park H et al (2001) Investigation on TiO_2-coated optical fibers for gas-phase photocatalytic oxidation of acetone. Appl Catal B 31(2001):209–220
Connel DW (2005) Basic concepts of environmental chemistry. Taylor and Francis, London

Dalrymple OK, Yeh DH, Trotz MA (2007) Removing pharmaceuticals and endocrine-disrupting compounds from wastewater by photocatalysis. J Chem Technol Biotechnol 82:121–134
De Padova P, Lucci M, Olivieri B et al (2009) Natural hybrid organic-inorganic photovoltaic devices. Superlattice Microst 45:555–563
Demeestere K, Dewulf J, De WB (2008) Heterogeneous photocatalytic removal of toluene from air on building materials enriched with TiO_2. Build Environ 43:406–414
Dholam R, Patel N, Adami M et al (2009) Hydrogen production by photocatalytic water-splitting using Cr- or Fe-doped TiO_2 composite thin films photocatalyst. Int J Hydrogen Energy 34:5337–5346
Ding L, Zhou H, Lou S et al (2013) Butterfly wing architecture assisted CdS/Au/TiO_2 Z-scheme type photocatalytic water splitting. Int J Hydrogen Energ 38:8244–8253
Dionysiou DD, Khodadoust AP, Kern AM et al (2000) Continuous-mode photocatalytic degradation of chlorinated phenols and pesticides in water using a bench-scale TiO_2 rotating disk reactor. Appl Catal B 24:139–155
Djebbar K, Sehili T (1998) Kinetics of heterogeneous photocatalytic decomposition of 2,4-dichlorophenoxyacetic acid over titanium dioxide and zinc oxide in aqueous solution. Pestic Sci 54:269–276
Doong R, Chen C-H, Maithreepala RA et al (2001) The influence of pH and cadmium sulphide on the photocatalytic degradation of 2-chlorophenol in titanium dioxide suspensions. Water Res 35:2873–2880
Driessen MD, Miller TM, Grassian VH (1998) Photocatalytic oxidation of trichloroethylene on zinc oxide: characterization of surface-bound and gas-phase products and intermediates with FT-IR spectroscopy. J Mol Catal A: Chem 131:149–156
Du Y, Zhou M, Lei L (2006) Role of the intermediates in the degradation of phenolic compounds by Fenton-like process. J Hazard Mater 136:859–865
Dutta PK, Pehkonen SO, Sharma VK et al (2005) Photocatalytic oxidation of arsenic (III): evidence of hydroxyl radicals. Environ Sci Technol 39:1827–1834
Einaga H, Futamura S, Ibusuki T (2001) Complete oxidation of benzene in gas phase by platinized titania photocatalysts. Environ Sci Technol 35:1880–1884
Epling GA, Lin C (2002) Photoassisted bleaching of dyes utilizing TiO_2 and visible light. Chemosphere 46:561–570
Evgenidou E, Konstantinou I, Fytianos K et al (2007) Photocatalytic oxidation of methyl parathion over TiO_2 and ZnO suspensions. Catal Today 124:156–162
Fateh R, Ismail AA, Dillert R et al (2011) Highly active crystalline mesoporous TiO_2 films coated onto polycarbonate substrates for self-cleaning applications. J Phys Chem C 115:10405–10411
Faust BC, Hoffmann MR, Bahnemann DW (1989) Photocatalytic oxidation of sulfur dioxide in aqueous suspensions of α-Fe_2O_3. J Phys Chem 93:6371–6381
Ferguson MA, Hering JG (2006) TiO_2-Photocatalyzed As(III) Oxidation in a fixed-bed, flow-through reactor. Environ Sci Technol 40:4261–4267
Fingler S, Dravenkar V (1988) Chlorophenols in the Sava river before, in and after the Zagreb City area. Impact on the purity of the city ground and drinking waters. Toxicol Environ Chem 17:319–328
Fox MA (1983) Organic heterogeneous photocatalysis: chemical conversions sensitized by irradiated semiconductors. Acc Chem Res 16:314–321
Fu X, Zeltner WA, Anderson MA (1995) The gas-phase photocatalytic mineralization benzene on porous titania-based catalysts. Appl Catal B 6:209–224
Fujishima K, Fukuoka A, Yamagishi A (2001) Photooxidation of benzene to phenol by ruthenium bipyridine complexes grafted on mesoporous silica FSM-16. J Mol Catal A: Chem 166:211–218
Gao J, Liu L, Liu X et al (2008) Levels and spatial distribution of chlorophenols—2,4-dichlorophenol, 2,4,6-trichlorophenol, and pentachlorophenol in surface water of China. Chemosphere 71:1181–1187
García A, Amat AM, Arques A et al (2006) Detoxification of aqueous solutions of the pesticide "Sevnol" by solar photocatalysis. Environ Chem Lett 3:169–172

García-López E, Marcí G, Serpone N et al (2007) Photoassisted oxidation of the recalcitrant cyanuric acid substrate in aqueous ZnO suspensions. J Phys Chem C 111:18025–18032

Gaya UI, Abdullah AH, Zainal Z et al (2009) Photocatalytic treatment of 4-chlorophenol in aqueous ZnO suspensions: intermediates, influence of dosage and inorganic anions. J Hazard Mater 168:57–63

Gaya UI, Abdullah AH, Zainal Z et al (2010) Photocatalytic degradation of 2,4-dichlorophenol in irradiated aqueous ZnO suspension. Int J Chem 2:180–193

Goel M, Chovelon J-M, Ferronato C et al (2010) The remediation of wastewater containing 4-chlorophenol using integrated photocatalytic and biological treatment. J Photochem Photobiol B 98:1–6

Grätzel M (2005) Solar energy conversion by dye-sensitized photovoltaic cells. Inorg Chem 44:6841–6851

Guillard C, Lachheb H, Houas A et al (2003) Influence of chemical structure of dyes, of pH and of inorganic salts on their photocatalytic degradation by TiO_2 comparison of the efficiency of powder and supported TiO_2. J Photochem Photobiol A 158:27–36

Haick H, Paz Y (2001) Remote photocatalytic activity as probed by measuring the degradation of self-assembled monolayers anchored near microdomains of titanium dioxide. J Phys Chem B 105:3045–3051

Hernández-Alonso MD, Coronado JM, Maira AJ et al (2002) Ozone enhanced activity of aqueous titanium dioxide suspensions for photocatalytic oxidation of free cyanide ions. Appl Catal B 39:257–267

Hernando MD, Mezcua M, Fernández-Alba AR et al (2006) Environmental risk assessment of pharmaceutical residues in wastewater effluents, surface waters and sediments. Talanta 69:334–342

Hiskia A, Ecke M, Troupis A et al (2001) Sonolytic, photolytic, and photocatalytic decomposition of atrazine in the presence of polyoxometalates. Environ Sci Technol 35:2358–2364

Hong S-S, Lee MS, Kim J-H et al (2002) Photocatalytic decomposition of bromate over titanium dioxides prepared using sol-gel method. J Ind Eng Chem 8:150–155

Horikoshi S, Hidaka H (2003) Non-degradable triazine substrates of atrazine and cyanuric acid hydrothermally and in supercritical water under the UV-illuminated photocatalytic cooperation. Chemosphere 51:139–142

Hsu C-W, Wang L, Su W-F (2009) Effect of chemical structure of interface modifier of TiO_2 on photovoltaic properties of poly(3-hexylthiophene)/TiO_2 layered solar cells. J Colloid Interface Sci 329:182–187

Huang H, Li W (2011) Destruction of toluene by ozone-enhanced photocatalysis: performance and mechanism. Appl Catal B 102:449–453

Huang C, Chen DH, Li K (2003) Photocatalytic oxidation of butyraldehyde over titania in air: by-product identification and reaction pathways. Chem Eng Comm 190:373–392

Huber R, Moser JE, Grätzel M et al (2002) Observation of photoinduced electron transfer in dye/semiconductor colloidal systems with different coupling strengths. Chem Phys 285:39–45

Hug SJ, Leupin O (2003) Iron-catalyzed oxidation of arsenic (III) by oxygen and by hydrogen peroxide: pH-dependent formation of oxidants in the fenton reaction. Environ Sci Technol 37:2734–2742

Hung C-H, Marias BJ (1997) Role of chlorine and oxygen in the photocatalytic degradation of trichloroethylene vapor on TiO_2 films. Environ Sci Technol 31:562–568

Ishibashi K-I, Nosaka Y, Hashimoto K et al (1998) Time-dependent behavior of active oxygen species formed on photoirradiated TiO_2 films in air. J Phys Chem B 102(12):2117–2120

Ishibashi K-I, Fujishima A, Watanabe T et al (2000) Generation and deactivation processes of superoxide formed on TiO_2 film illuminated by very weak UV light in air or water. J Phys Chem B 104:4934–4938

Jacoby WA, Blake DM, Fennell JA (1996) Heterogeneous photocatalysis for control of volatile organic compounds in indoor air. J Air Waste Manage Assoc 46:891–898

Jańczyk A, Krakowska E, Stochel G (2006) Singlet oxygen photogeneration at surface modified titanium dioxide. J Am Chem Soc 128:15574–15575

Jardim WF, Moraes SG, Takiyama MMK (1997) Photocatalytic degradation of aromatic chlorinated compounds using TiO_2: toxicity of intermediates. Wat Res 31:1728–1732

Jayaweera PVV, Perera AGU, Tennakone K (2008) Why Gratzel's cell works so well. Inorg Chim Acta 361:707–711

Jiang Y, Petrier C, Waite TD (2006) Sonolysis of 4-chlorophenol in aqueous solution: effects of substrate concentration, aqueous temperature and ultrasonic frequency. Ultrason Sonochem 13:415–422

Jung KY, Park SB (2004) Photoactivity of SiO_2/TiO_2 and ZrO_2/TiO_2 mixed oxides prepared by sol-gel method. Mater Lett 58:2897–2900

Kamble SP, Deosarkar SP, Sawant SB (2004) Photocatalytic Degradation of 2,4-Dichlorophenoxyacetic Acid Using Concentrated Solar Radiation: Batch and Continuous Operation. Ind Eng Chem Res 43(26):8178–8187

Kandiel TA, Dillert R, Robben L et al (2011) Photonic efficiency and mechanism of photocatalytic molecular hydrogen production over platinized titanium dioxide from aqueous methanol solutions. Catal Today 161:196–201

Kebir M, Chabani M, Nasrallah N et al (2011) Coupling adsorption with photocatalysis process for the Cr(VI) removal. Desalination 270:166–173

Khodja AA, Sehili T, Pilichowski J-F et al (2001) Photocatalytic degradation of 2-phenylphenol on TiO_2 and ZnO in aqueous suspensions. J Photochem Photobiol A 141:231–239

Kim H, Choi W (2007) Effects of surface fluorination of TiO_2 on photocatalytic oxidation of gaseous acetaldehyde. Appl Catal B 69:127–132

Kong F-T, Dai S-Y, Wang K-J (2007) Review of recent progress in dye-sensitized solar cells. Adv OptoElectron. doi:10.1155/2007/75384

Konstantinou IK, Sakkas VA, Albanis TA (2001) Photocatalytic degradation of the herbicides propanil and molinate over aqueous TiO_2 suspensions: identification of intermediates and the reaction pathway. Appl Catal B 34:227–239

Krijgsheld KR, Gen A van der (1986) Assessment of the impact of the emission of certain organochlorine compounds on the aquatic environment. Chemosphere 15:825–860

Kubo W, Tatsuma T (2004) Detection of H_2O_2 released from TiO_2 phototcatalyst to air. Anal Sci 24:591–593

Kubo W, Tatsuma T (2005) Photocatalytic remote oxidation with various photocatalysts and enhancement of its activity. J Mater Chem 15:3104–3108

Kubo W, Tatsuma T, Fujishima A et al (2004) Mechanisms and resolution of photocatalytic lithography. J Phys Chem B 108:3005–3009

Kwon BG, Yoon J (2009) Experimental evidence of the mobility of hydroperoxyl/superoxide anion radicals from the illuminated TiO_2 interface into the aqueous phase. Bull Korean Chem Soc 30(3):667–670

Kyung HS, Lee JS, Choi W (2005) Simultaneous and synergistic conversion of dyes and heavy metal ions in aqueous TiO_2 suspensions under visible-light illumination. Environ Sci Technol 39:2376–2382

L'Amour RJA, Azevedo EB, Leite SGF et al (2008) Removal of phenol in high salinity media by a hybrid process (activated sludge + photocatalysis). Sep Purif Technol 60:142–146

L'homme L, Brosillon S, Wolbert D (2008) Photocatalytic degradation of pesticides in pure water and a commercial agricultural solution on TiO_2 coated media. Chemosphere 70(2008):381–386

Lackhoff M, Niessner R (2002) Photocatalytic atrazine degradation by synthetic minerals, atmospheric aerosols, and soil particles. Environ Sci Technol 36:5342–5347

Lai K, Zhu Y, Lu J et al (2013) N- and Mo-doping Bi_2WO_6 in photocatalytic water splitting. Comput Mater Sci 67:88–92

Lalitha K, Reddy JK, Sharma MVP et al (2010) Continuous hydrogen production activity over finely dispersed Ag_2O/TiO_2 catalysts from methanol:water mixtures under solar irradiation: a structure–activity correlation. Int J Hydrogen Energy 35:3991–4001

Lathasree S, Rao AN, SivaSankar B et al (2004) Heterogeneous photocatalytic mineralisation of phenols in aqueous solutions. J Mol Catal A: Chem 223:101–105

Lee H, Choi W (2002a) Photocatalytic oxidation of arsenite in TiO_2 suspension: kinetics and mechanisms. Environ Sci Technol 36:3872–3878

Lee MC, Choi W (2002b) Solid phase photocatalytic reaction on the soot/TiO_2 interface: the role of migrating OH radicals. J Phys Chem B 106:11818–11822

Leng WH, Cheng XF, Zhang JQ et al (2007) Comment on "Photocatalytic oxidation of arsenite on TiO_2: understanding the controversial oxidation mechanism involving superoxides and the effect of alternative electron acceptors". Environ Sci Technol 41:6311–6312

Lewerenz HJ, Heine C, Skorupska K et al (2010) Photoelectrocatalysis: principles, nanoemitter applications and routes to bio-inspired systems. Energy Environ Sci 3:748–760

Li D, Haneda H (2003) Photocatalysis of sprayed nitrogen-containing Fe_2O_3–ZnO and WO_3–ZnO composite powders in gas-phase acetaldehyde decomposition. J Photochem Photobiol A 160:203–212

Li S, Ma Z, Zhang J et al (2008) A comparative study of photocatalytic degradation of phenol of TiO_2 and ZnO in the presence of manganese dioxides. Catal Today 139:109–112

Li Q, Easter NJ, Shang JK (2009) As(III) removal by palladium-modified nitrogen-doped titanium oxide nanoparticle photocatalyst. Environ Sci Technol 43:1534–1539

Li W, Lu S, Qiu Z et al (2011) Photocatalysis of clofibric acid under solar light in summer and winter seasons. Ind Eng Chem Res 50:5384–5393

Liqiang J, Baifu X, Fulong Y et al (2004) Deactivation and regeneration of ZnO and TiO_2 nanoparticles in the gas phase photocatalytic oxidation of n-C_7H_{16} or SO_2. Appl Catal A 275:49–54

Luo Y, Ollis DF (1996) Heterogeneous photocatalytic oxidation of trichloroethylene and toluene mixtures in air: kinetic promotion and inhibition, time-dependent catalyst activity. J Catal 163:1–11

Maira AJ, Yeung KL, Soria J et al (2001) Gas-phase photo-oxidation of toluene using nanometer-size TiO_2 catalysts. Appl Catal B 29:327–336

Malato S, Cáceres J, Fernández-Alba AR (2003) Photocatalytic treatment of diuron by solar photocatalysis: evaluation of main intermediates and toxicity. Environ Sci Technol 37:2516–2524

Mantilla A, Tzompantzi F, Fernández JL et al (2010) Photodegradation of phenol and cresol in aqueous medium by using Zn/Al + Fe mixed oxides obtained from layered double hydroxides materials. Catal Today 150:353–357

Marchante E, Lana-Villarreal T, Sáez V, Gonzólez-García J, Gómez R (2009) Sonopotential: a new concept in electrochemistry. Chem Commun 27:4127–4129

Martra G, Coluccia S, Marchese L (1999) The role of H_2O in the photocatalytic oxidation of toluene in vapour phase on anatase TiO_2 catalyst. A FTIR study. Catal Today 53:695–702

Matar S, Hatch LF (2000) Chemistry of petrochemical processes. Gulf Publishing Company, Texas

McCusker JK (2001) Fuel from photons. Science 293:1599–1601

McMurray TA, Dunlop PSM, Byrne JA (2006) The photocatalytic degradation of atrazine on nanoparticulate TiO_2 films. J Photochem Photobiol A 182:43–51

Mills A, McGrady M (2008) A study of new photocatalyst indicator inks. J Photochem Photobiol A 193:228–236

Mills A, Belghazi A, Rodman D (1996) Bromate removal from water by semiconductor photocatalysis. Water Res 30:1973–1978

Mills A, Hill G, Bhopal S et al (2003) Thick titanium dioxide films for semiconductor photocatalysis. J Photochem Photobiol A 160:185–194

Mills A, Wang J, Lee S-K et al (2005) An intelligence ink for photocatalytic films. Chem Commun 21:2721–2723

Mills A, Wang J, McGrady M (2006) Method of rapid assessment of photocatalytic activities of self-cleaning films. J Phys Chem B 110:18324–18331

Mills A, McGrady M, Wang J et al (2008) A rapid method of assessing the photocatalytic activity of thin TiO_2 films using an ink based on the redox dye 2,6-dichloroindophenol. Int J Photoenergy: 1–6

Minero C, Mariella G, Maurino V et al (2000a) Photocatalytic transformation of organic compounds in the presence of inorganic anions. 1. Hydroxyl-mediated and direct electron-transfer reactions of phenol on a titanium dioxide-fluoride system. Langmuir 16:2632–2641

Minero C, Mariella G, Maurino V et al (2000b) Photocatalytic transformation of organic compounds in the presence of inorganic ions. 2. Competitive reactions of phenol and alcohols on a titanium dioxide-fluoride system. Langmuir 16:8964–8972

Minero C, Pelizzetti E, Malato S et al (1996) Large solar plant photocatalytic water decontamination: degradation of atrazine. Sol Energy 56:411–419

Mofidi AA, Min JH, Palencia LS et al (2002) Task 2.1: advanced oxidation processes and UV Photolysis for treatment of drinking water, metropolitan water district of Southern California. California Energy Commission, La Verne

Mohan D, Pittman CU Jr (2007) Arsenic removal from water/wastewater using adsorbents—a critical review. J Hazard Mater 142:1–53

Moser J, Grätzel M (1983) Light-induced electron transfer in colloidal semiconductor dispersions: single vs. dielectronic reduction of acceptors by conduction-band electrons. J Am Chem Soc 105:6547–6555

Muradov NZ, Raissi AT, Muzzey D et al (1996) Selective photocatalytic destruction of airbone VOCs. Sol Energy 56:445–456

Murakami Y, Kenji E, Nosaka AY et al (2006) Direct detection of OH radicals diffused to the gas phase from the UV-irradiated photocatalytic TiO_2 surfaces by means of laser-induced fluorescence spectroscopy. J Phys Chem B 110(34):16808–16811

Murakami Y, Endo K, Ohta I et al (2007) Can OH radicals diffuse from the UV-irradiated photocatalytic TiO_2 surfaces? Laser-induced-fluorescence study. J Phys Chem C 111:11339–11346

Nagata Y, Nakagawa M, Okuno H et al (2000) Sonochemical degradation of chlorophenols in water. Ultrason Sonochem 7:115–120

Naito K, Tachikawa T, Cui S-C et al (2006) Single-molecule detection of airborne singlet oxygen. J Am Chem Soc 128:16430–16431

Nakamura R, Nakato Y (2004) Primary intermediates of oxygen photoevolution reaction on TiO_2 (rutile) particles, revealed by in situ FTIR absorption and photoluminescence measurements. J Am Chem Soc 126:1290–1298

Neta P, Huie RE (1985) Free-radical chemistry of sulphite. Environ Health Perspect 64:209–217

Nikolaou AD, Kostopoulou MN, Lekkas TD (1999) Organic by-products of drinking water chlorination. Global Nest: Int J 1:143–156

Nimlos MR, Jacoby WA, Blake DM et al (1979) Direct mass spectrometric studies of the destruction of hazardous wastes. 2. Gas-phase photocatalytic oxidation of trichloroethylene over TiO_2: Products and mechanisms. Environ Sci Technol 27:732–740

Ohko Y, Tryk DA, Hashimoto K et al (1998) Autoxidation of acetaldehyde initiated by TiO_2 photocatalysis under weak UV illumination. J Phys Chem B 102:2699–2704

Ojanperä I (2000) Pesticides. In: Bogusz MJ (ed) Forensic science. Hanbook of analytical separations, vol 2. Elsevier Science

Oliveira HG, Nery DC, Longo C (2010) Effect of applied potential on photocatalytic phenol degradation using nanocrystalline TiO_2 electrodes. Appl Catal B 93:205–211

Oturan N, Brillas E, Oturan MA (2012) Unprecedented total mineralization of atrazine and cyanuric acid by anodic oxidation and electro-Fenton with a boron-doped diamond anode. Environ Chem Lett 10:165–170

Özcan L, Yurdakal S, Augugliaro V et al (2010) Photoelectrocatalytic selective oxidation of 4-methoxybenzyl alcohol in water by TiO_2 supported on titanium anodes. Appl Catal B. doi:10.1016/j.apcatb.2012.12.030

Park H, Choi W (2004) Effects of TiO_2 surface fluorination on photocatalytic reactions and photoelectrochemical behaviours. J Phys Chem B 108:4086–4093

Parra S, Stanca SE, Guasaquillo I et al (2004) Photocatalytic degradation of atrazine using suspended and supported TiO2. Appl Catal B 51:107–116

Patnaik P (2004) DeanÊ¼s analytical chemistry handbook. McGraw-Hill Handbooks, United States of America

Paul T, Miller PL, Strathmann TJ (2007) Visible-light-mediated TiO_2 photocatalysis of fluoroquinolone antibacterial agents. Environ Sci Technol 41:4720–4727

Pelizetti E, Maurino V, Minero C (1990) Photocatalytic degradation of atrazine and other s-triazine herbicides. Environ Sci Technol 24:1559–1565

Peller J, Wiest O, Kamat PV (2003) Synergy of combining sonolysis and photocatalysis in the degradation and mineralization of chlorinated aromatic compounds. Environ Sci Technol 37:1926–1932

Peller J, Wiest O, Kamat PV (2004) Hydroxyl radical's role in the remediation of a common herbicide, 2,4-Dichlorophenoxyacetic acid (2,4-D). J Phys Chem A 108:10925–10933

Pena ME, Korfiatis GP, Patel M et al (2005) Adsorption of As(V) and As(III) by nanocrystalline titanium dioxide. Water Res 39:2327–2337

Peralta-Zamora P, Gomes deMS, Pelegrinir R et al (1998) Evaluation of ZnO, TiO_2 and supported ZnO on the photoassisted remediatlon of black liquor, cellulose and textile mill effluents. Chemosphere 36:2119–2133

Pichat P, Herrmann J-M, Disdler J et al (1979) Photocatalytic oxidation of propene over various oxides at 320 K. Selectivity. J Phys Chem 83:3122–3126

Poulose AC, Veeranarayanan S, Varghese SH et al (2012) Functionalized electrophoretic deposition of CdSe quantum dots onto TiO_2 electrode for photovoltaic application. Chem Phys Lett 539–540:197–203

Prairie MR, Evans LR, Stange BM et al (1993) An investigation of TiO_2 photocatalysis for the treatment of water contaminated with metals and organic chemicals. Environ Sci Technol 27:1776–1782

Qamar M, Muneer M, Bahnemann D (2006) Heterogeneous photocatalysed degradation of two selected pesticide derivatives, triclopyr and daminozid in aqueous suspensions of titanium dioxide. J Environ Manage 80:99–106

Rahman MA, Muneer M (2005) Photocatalysed degradation of two selected pesticide derivatives, dichlorvos and phosphamidon, triclopyr and daminozid in aqueous suspensions of titanium dioxide. Desalination 185:161–172

Ranjit PJD, Palanivelu K, Lee C-S (2008) Degradation of 2,4-dichlorophenol in aqueous solution by sono-Fenton method. Korean J Chem Eng 25:112–117

Rauf MA, Meetani MA, Hisaindee S (2011) An overview on the photocatalytic degradation of azo dyes in the presence of TiO_2 doped with selective transition metals. Desalination 276:13–27

Ravichandran L, Selvam K, Swaminathan M (2007) Effect of oxidants and metal ions on photodefluoridation of pentafluorobenzoic acid with ZnO. Sep Purif Methods 56:192–198

Rayalu SS, Jose D, Joshi MV et al (2013) Photocatalytic water splitting on Au/TiO_2 nanocompositessynthesized through various routes: enhancement in photocatalyticactivity due to SPR effect. Appl Catal B 142–143:684–693

Razykov TM, Ferekides CS, Morel D et al (2011) Solar photovoltaic electricity: current status and future prospects. Sol Energy 85:1580–1608

Reyes C, Fernandez J, Freer J et al (2006) Degradation and inactivation of tetracycline by TiO_2 photocatalysis. J Photochem Photobiol A 184:141–146

Ryu J, Choi W (2004) Effects of TiO_2 surface modifications on photocatalytic oxidation of arsenite: the role of superoxides. Environ Sci Technol 38:2928–2933

Ryu J, Choi W (2006) Photocatalytic oxidation of arsenite on TiO_2: Understanding the controversial oxidation mechanism involving superoxides and the effect of alternative electron acceptors. Environ Sci Technol 40:7034–7039

Ryu J, Choi W (2007) Response to comment on "Photocatalytic oxidation of arsenite on TiO_2: understanding the controversial oxidation mechanism involving superoxides and the effect of alternative electron acceptors". Environ Sci Technol 41:6313–6314

Sakthivel S, Neppolian B, Shankar MV et al (2003) Solar photocatalytic degradation of azo dye: comparison of photocatalytic efficiency of ZnO and TiO_2. Sol Energy Mater Sol Cells 77:65–82
Sandstead HH (1991) Zinc deficiency. A public health problem? Am J Dis Child 145(8):853–859
Sasaki Y, Iwase A, Kato H et al (2008) The effect of co-catalyst for Z-scheme photocatalysis systems with an Fe^{3+}/Fe^{2+} electron mediator on overall water splitting under visible light irradiation. J Catal 259(2008):133–137
Sauer ML, Hale MA, Ollis DF (1995) Heterogeneous photocatalytic oxidation of dilute toluene-chlorocarbon mixtures in air. J Photochem Photobiol A 88:169–178
Sawyer DT, Sobkowiak A, Roberts JL Jr (1995) Electrochemistry for chemists. Wiley, New York
Sharma VK, Sohn M (2009) Aquatic arsenic: toxicity, speciation, transformations, and remediation. Environ Int 35:743–759
Shen YS, Ku Y (2002) Decomposition of gas-phase trichloroethene by the UV/TiO_2 process in the presence of ozone. Chemosphere 46:101–107
Shen P, Liu X, Jiang S et al (2011) Effects of aromatic p-conjugated bridges on optical and photovoltaic properties of N, N-diphenylhydrazone-based metal-free organic dyes, N, N-diphenylhydrazone-based metal-free organic dyes. Org Electron 12:1992–2002
Simamora A-J, Hsiung T-L, Chang F-C et al (2013) Photocatalytic splitting of seawater and degradation of methylene blue on CuO/nano TiO_2. Int J Hydrogen Energy 37:13855–13858
Sleiman M, Ferronato C, Chovelon J-M (2008) Photocatalytic removal of pesticide dichlorvos from indoor air: a study of reaction parameters, intermediates and mineralization. Environ Sci Technol 42:3018–3024
Sobana N, Swaminathan M (2007) The effect of operational parameters on the photocatalytic degradation of acid red 18 by ZnO. Sep Purif Technol 56:101–107
Somasundaram S, Chenthamarakshan CRN, de Tacconi NR et al (2007) Photocatalytic production of hydrogen from electrodeposited p-Cu_2O film and sacrificial electron donors. Int J Hydrogen Energ 32:4661–4669
Stock NL, Peller J, Vinodgopal K et al (2000) Combinative sonolysis and photocatalysis for textile dye degradation. Environ Sci Technol 34:1747–1750
Sun L, Bolton JR (1996) Determination of the quantum yield for the photochemical generation of hydroxyl radicals in TiO_2 suspensions. J Phys Chem 100:4127–4134
Sun B, Reddy EP, Smirniotis PG (2005) Visible light Cr(VI) reduction and organic chemical oxidation by TiO_2 photocatalysis. Environ Sci Technol 39:6251–6259
Suryaman D, Hasegawa K, Kagaya S (2006) Combined biological and photocatalytic treatment for the mineralization of phenol in water. Chemosphere 65:2502–2506
Swaddle TW (1990) Inorganic chemistry. Academic, USA
Tatsuma T, Kubo W (2007) Photocatalytic lithography based on photocatalytic remote oxidation. J Photopolym Sci Technol 20:83–87
Tatsuma T, Tachibana S-I, Miwa T et al (1999) Remote bleaching of methylene blue by UV-irradiated TiO_2 in the gas phase. J Phys Chem B 103:8033–8035
Tatsuma T, Kubo W, Fujishima A (2002) Patterning of solid surfaces by photocatalytic lithography based on the remote oxidation effect of TiO_2. Langmuir 18:9632–9634
Tetzlaff TA, Jenks WS (1999) Stability of cyanuric acid to photocatalytic degradation. Org Lett 1:463–465
Thampi KR, Reddy TV, Ramakrishnan V et al (1983) Mechanism of the photoelectrocatalytic dehydrogenation of 2-propanolon a polycrystalline ZnO photoelectrode. Electrochim Acta 28:1869–1874
Turchi CS, Ollis DF (1990) Photocatalytic degradation of organic water contaminants: mechanisms involving hydroxyl radical attack. J Catal 122:178–192
Uddin MJ, Davies B, Dickens TJ et al (2013) Self-aligned carbon nanotubes yarns (CNY) with efficient optoelectronic interface for microyarn shaped 3D photovoltaic cells. Sol Energy Mater Sol Cells 115:166–171

US EPA (2006) National Emission Standards for Hazardous Air Pollutants: miscellaneous organic chemical manufacturing. final rule, federal register part V, 40 CFR part 63, United States Environmental Protection Agency

EPA/OSHA US (2000) Chemical advisory and notice of potential risk: skin exposure to molten 2,4-dichlorophenol (2,4-DCP) can cause rapid death, US EPA Office of pollution prevention and toxics and occupational safety and health administration, United States of America

Vimonses V, Jin B, Chow CWK et al (2010) An adsorption-photocatalysis hybrid process using multi-functional-nanoporous materials for wastewater reclamation. Water Res 44:5385–5397

Vincent G, Aluculesei A, Parker A et al (2008) Direct detection of OH radicals and indirect detection of H_2O_2 molecules in the gas phase near a TiO_2 photocatalyst using LIF. J Phys Chem C 112(25):9115–9119

Vinu R, Madras G (2009) Kinetics of sonophotocatalytic degradation of anionic dyes with nano-TiO_2. Environ Sci Technol 43:473–479

Vorontsov AV, Stoyanova IV, Kozlov DV et al (2000) Kinetics of the photocatalytic oxidation of gaseous acetone over platinized titanium dioxide. J Catal 189:360–369

Wang K-H, Jehng J-M, Hsieh Y-H et al (2002) The reaction pathway for the heterogeneous photocatalysis of trichloroethylene in gas phase. J Hazard Mater 90:63–75

Wang X, Jia J, Wang Y (2011) Degradation of C. I. reactive red 2 through photocatalysis coupled with water jet cavitation. J Hazard Mater 185:315–321

Wang Q, An N, Bai Y et al (2013) High photocatalytic hydrogen production from methanol aqueous solution using the photocatalysts CuS/TiO_2. Int J Hydrogen Energy. http://dx.doi.org/10.1016/j.ijhydene.2013.02.131

Wegman RCC, Den BHH van (1983) Chlorophenols in river sediments in the Netherlands (1976–1977). Water Res 17:227–230

Wu W, Li J, Guo F et al (2010) Photovoltaic performance and long-term stability of quasi-solid-state fluoranthene dyes-sensitized solar cells. Renew Energ 35:1724–1728

Wunderlich W, Oekermann T, Miao L et al (2004) Electronic properties of nano-porous TiO_2- and ZnO-thin films-comparison of simulations and experiments. J Ceram Process Res 5:343–354

Xu T, Kamat PV, O'Shea KE (2005) Mechanistic evaluation of arsenite oxidation in TiO_2 assisted photocatalysis. J Phys Chem A 109:9070–9075

Yahiat S, Fourcade F, Brosillon S et al (2011a) Photocatalysis as a pre-treatment prior to a biological degradation of cyproconazole. Desalination 281:61–67

Yahiat S, Fourcade F, Brosillon S et al (2011b) Removal of antibiotics by an integrated process coupling photocatalysis and biological treatment—case of tetracycline and tylosin. Int Biodeter and Biodegr 65:997–1003

Yamaguchi M, Tomizawa M, Takagaki K, Masui D, Yamagishi T et al (2006) Photooxidation of alkane under visible light irradiation catalyzed by ruthenium complexes. Catal Today 117:206–209

Yan J, Zhang L, Yang H et al (2009) $CuCr_2O_4/TiO_2$ heterojunction for photocatalytic H_2 evolution under simulated sunlight irradiation. Sol Energy 83:1534–1539

Yang H, Lin WY, Rajeshwar K (1999) Homogeneous and heterogeneous photocatalytic reactions involving As(III) and As(V) species in aqueous media. J Photochem Photobiol A 123:137–143

Yang G-P, Zhao X-K, Suna X-J et al (2005a) Oxidative degradation of diethyl phthalate by photochemically-enhanced Fenton reaction. J Hazard Mater 126:112–118

Yang J, Chen C, Ji H et al (2005b) Mechanism of TiO_2-assisted photocatalytic degradation of dyes under visible irradiation: photoelectrocatalytic study by TiO_2-film Electrodes. J Phys Chem B 109:21900–21907

Yang L, Yu LE, Ray MB (2009) Photocatalytic oxidation of paracetamol: dominant reactants, intermediates, and reaction mechanisms. Environ Sci Technol 43:460–465

Yoon SH, Lee JH (2005) Oxidation mechanism of As(III) in the UV/TiO_2 system: evidence for a direct hole oxidation mechanism. Environ Sci Technol 39:9695–9701

Yoon SH, SE Oh, Yang JE et al (2009) TiO2 photocatalytic oxidation mechanism of As(III). Environ Sci Technol 43:864–869

Zainal Z, Lee CY, Hussein MZ et al (2007) Electrochemical-assisted photodegradation of mixed dye and textile effluents using TiO_2 thin films. J Hazard Mater 146:73–80
Zainudin NF, Abdullah AZ, Mohamed AR (2010) Characteristics of supported nano-TiO_2/ZSM-5/silica gel (SNTZS): photocatalytic degradation of phenol. J Hazard Mater 174:299–306
Zama K, Fukuoka A, Sasaki Y (2000) Selective hydroxylation of benzene to phenol by photocatalysis of molybdenum complexes grafted on mesoporous FSM-16. Catal Lett 66:251–253
Zhang F, Zhaoa J, Shen T et al (1998) TiO_2-assisted photodegradation of dye pollutants II. Adsorption and degradation kinetics of eosin in TiO_2 dispersions under visible light irradiation. Appl Catal B 15:147–156
Zhang Q, Liu J, Wang W et al (2006a) A novel regular catalyst for gas-phase photodestruction of organic substances of dilute concentrations. Catal Commun 7:685–688
Zhang S, Zheng Z, Wang J et al (2006b) Heterogeneous photocatalytic decomposition of benzene on lanthanum-doped TiO_2 film at ambient temperature. Chemosphere 65:2282–2288
Zheng S, Cai Y, O'Shea KE (2010) TiO_2 photocatalytic degradation of phenylarsonic acid. J Photochem Photobiol A 210:61–68
Zhong J, Wang J, Tao L et al (2007) Photocatalytic degradation of gaseous benzene over TiO_2/Sr_2CeO_4: kinetic model and degradation mechanisms. J Hazard Mater 139:323–331
Zhu X, Yuan CW, Chen H (2007) Photocatalytic degradation of pesticide pyridaben. 3. In surfactant/TiO_2 aqueous dispersions. Environ Sci Technol 41:263–269
Zou S-W, How C-W, Chen JP (2007) Photocatalytic treatment of wastewater contaminated with organic waste and copper ions from the semiconductor industry. Ind Eng Chem Res 46:6566–6571
Zuo G-M, Cheng Z-X, Chen H (2006) Study on photocatalytic degradation of several volatile organic compounds. J Hazard Mater 128:158–163

Chapter 6
Photosynthetic Routes to Organic Compounds

Abstract The potentials of semiconductor photocatalysis in organic synthesis cannot be overemphasised. By far, many organic substances have been successfully synthesised at laboratory scale. Some revolutionary features of these photocatalytic routes include greenness and the ability to proceed in one pot. Herein, an overview of the routes and suggested mechanisms to important functional groups is given, and where necessary, contrast with the conventional routes is made.

The potentials of semiconductor photocatalysis in organic synthesis cannot be overemphasised. By far, many organic substances have been successfully synthesised at laboratory scale. Some revolutionary features of these photocatalytic routes include greenness and the ability to proceed in one pot. Herein, an overview of the routes and suggested mechanisms to important functional groups is given, and where necessary, contrast with the conventional routes is made.

6.1 Background Sketch

Thermal heterogeneous photocatalytic synthesis is a given, and forms the basis of many known industrial processes. However, owing to the extreme reaction conditions involved in thermal processes they are usually costly and require special metallurgy for operation. Photocatalytic synthesis contrasts sharply with conventional synthesis routes for the fact that it can proceed successfully at ambient conditions. Basically, reactions on the surface of semiconductor photocatalyst can be designed to proceed by either photosynthesis or photodegradation pathway. Mechanistically, photocatalytic synthesis differs from photocatalytic degradation in many respects. Whereas photocatalytic degradation usually proceeds non-selectively, photocatalytic synthesis usually proceeds via selective organic reactions. One clear difference is that photocatalytic synthesis can proceed favourably under, partially hydrated or neat organic solvent condition depending upon the target product. In fact, under dearated conditions photocatalytic reduction (such as alcohol reduction) by electrons can proceed independent of hole reaction.

Of the characteristics of semiconductor photocatalysts that concern photocatalytic synthesis, selectivity and turnover frequency are crucial. By definition, selectivity for a given compound or functionality refers to the formation of the chemical compound

U. I. Gaya, *Heterogeneous Photocatalysis Using Inorganic Semiconductor Solids*, DOI 10.1007/978-94-007-7775-0_6,

or fuctionality in preference to other chemical compounds or functionalities. Even within the same photocatalytic system different products can be formed. The choice of catalyst helps in improving the selectivity of a desired reaction. Turnover number on the other hand, can be defined as the number of moles of products produced per mole of the catalyst, and this parameter becomes turnover frequency when estimated per unit time. Even though, semiconductor metal oxides can sometimes act as high-temperature thermal oxidation catalysts, better oxidative selectivity is often observed in the room temperature photocatalytic process (Fox and Dulay 1995).

Ideas on phototocatalytic synthesis have recently flourished, covering important areas in organic chemistry such as cyclisation and functional group conversion. In some many cases, these ideas are based on the chemistry developed from known homogeneous or heterogeneous reactions that are associated with the generation of hydroxyl radicals. Kolbe-electrolysis and photolysis are good examples to support this fact. Nano-sized semiconductor materials are currently attracting interest as powerful photocatalysts for exploitation in organic synthesis. Generally speaking, the synthesis of organic compounds by heterogeneous processes offers superior advantage over homogeneous counterparts due to the elimination of need to separate the catalyst in the final stages. By far, extensive research activities have been carried out with special focus on the practical application of photocatalysis in organic synthesis. In this chapter, we highlight some important photosynthetic reactions achieved over semiconductor photocatalysts such as photocatalytic epoxidation, oxidation of alkanes, oxidation of aromatic compounds, methane conversion, alcohol conversion and coupling of methane.

6.2 Reductive Routes

6.2.1 Photo-Kolbe Reduction

Kolbe electrolysis is a classical, great reduction method of production of radical species for electroorganic synthesis. These reactions were first studied in 1849 by Hermann Kolbe, who demonstrated the decarboxylation of acetate using electrolysis. The Kolbe reactions are shown by Eq. (6.1)–(6.3).

$$2CH_3COO^- \rightarrow 2CH_3COO^\bullet + 2e^- \tag{6.1}$$

$$2CH_3COO^\bullet \rightarrow 2CH_3{}^\bullet + 2CO_2 \tag{6.2}$$

$$2CH_3{}^\bullet \rightarrow C_2H_6 \tag{6.3}$$

The photocatalytic version of the Kolbe synthesis was for the first time demonstrated by Kraeutler and Bard (Kreutler and Bard 1978) using Pt/TiO_2 as a photocatalyst for the reduction of ethanoic acid to methane and CO_2. This reaction may be represented as follows:

$$CH_3COOH \rightarrow CH_4 + CO_2 \tag{6.4}$$

The decomposition of other saturated alkanoic acids proceeds in the same way. Therefore, we may generally use Eq. (6.5) as equation for photocatalytic Kolbe conversion:

$$RCO_2H \xrightarrow{h\nu} RH + CO_2 \tag{6.5}$$

The Kolbe photocatalytic reduction of propanoic acid, butanoic and pentanoic acid may be represented by the following major reactions (Kreutler and Bard 1978):

$$CH_3CH_2CO_2H \rightarrow CH_3CH_3 + CO_2 \tag{6.6}$$

$$CH_3CH_2CH_2CO_2H \rightarrow CH_3CH_2CH_3 + CO_2 \tag{6.7}$$

$$CH_3CH_2CH_2CH_2CO_2H \rightarrow CH_3CH_2CH_2CH_3 + CO_2 \tag{6.8}$$

In most of these decarboxylative processes, small amounts of by-products are also formed which may vary depending upon the starting material or the presence of oxygen. In the conversion of ethanoic acid, the minor products are H_2 and ethane (5 to 10 %).

$$2CH_3COOH \rightarrow C_2H_6 + 2CO_2 + H_2 \tag{6.9}$$

Kraeutler and Bard utilised liquid carboxylic acids for the photocatalytic Kolbe reaction. Subsequently, Sato (1982) disclosed the synthesis of ethane from gaseous acetic acid, in a platinised TiO_2 system. The products of this gas-solid interfacial reaction were dependent upon the reaction conditions. For example, remarkable increase in decarboxylation rate was observed after the addition of gaseous H_2O.

6.2.2 *Reduction of Nitroaromatic Compounds*

The photocatalytic reduction of nitro-functional group has been demonstrated despite the fact that not much studies have been reported in this context. Wu et al. (2013) investigated the photocatalytic conversion of *p*-nitroaniline to *p*-phenylenediamine over TiO_2, ZnO, $PbBi_2Nb_2O_9$ and CdS in the presence of methanol. As may be expected, TiO_2 and ZnO showed photocatalytic activities for the photoreduction of 4-nitroaniline. Electron spin resonance analysis revealed $^{\bullet}CH_2OH$ in the presence of TiO_2 and ZnO. The formation of this radical was attributed to the reaction between hydroxyl radical and CH_3OH molecule. Both species could reduce 4-nitroaniline to *p*-phenylenediamine. Other alcohols such as C_2H_5OH and *i*-C_3H_7OH were also found to be efficient additives for the photoreduction of the compound 4-nitroaniline. Evidences from ^2H (D)-labeled experiments indicated that the *p*-phenylenediamine formation involves hydrogen transfer reaction between the *p*-nitroaniline and H_2O molecules as shown in Fig. 6.1.

In a different study, the mechanism of reduction of nitroorganics has been extensively investigated. The photocatalytic reduction of array of mono-suubstituted nitrobenzenes (4-nitrotoluene, 3-nitrotoluene, 4-nitrophenol, 3-nitrophenol, 4-nitrobenzonitrile, 3-nitrobenzonitrile, 4-(R, R,R-trifluoromethyl)nitrobenzene, 3-(R,

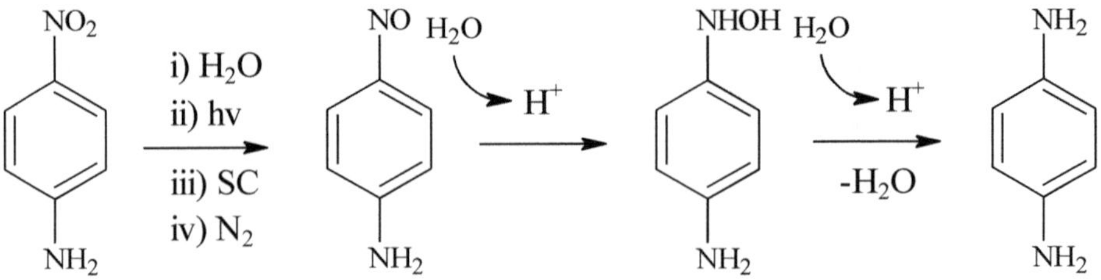

Fig. 6.1 Reaction scheme for the reductive conversion of *p*-nitroaniline to *p*-phenylenediamine

R,R-trifluoromethyl)-nitrobenzene, 4-chloronitrobenzene, 3-chloronitrobenzene, 4-aminotoluene, 3-aminotoluene, 4-aminobenzonitrile, 3-aminobenzonitrile, nitrosobenzene, aniline, tribromomethane, tetranitromethane) in deoxygenated, illuminated, aqueous suspensions of P25 titanium dioxide in the presence of 2-propanol as sacrificial electron donor, agreed remarkably with Langmuir-Hinshelwood scheme (Ferry and Glaze 1998). The highest rate constant of reduction (9×10^7 mol dm^{-3}s^{-1}) was obtained with nitrobenzonitrile as starting material.

6.3 Partial Alcohol Oxidation

Many industrially important precursors are synthesised by oxidative processes. Carbonyl compounds such as aldehydes and ketones are used as intermediaries in the synthesis of drugs, fine chemicals and fragrances. For example, one of the processes that produce ethyl ethanoate, an important industrial solvent, has been until recently the Tischenko reaction of ethanal (Jackson et al. 2003) which is catalysed by aluminium alkoxide. The photocatalytic reduction of aliphatic alcohols has been studied in detail by Nishimoto et al. (1985). The wavelength threshold for effective photoconversion in deareated, aqueous platinised TiO_2 is < 390 nm. All the alcohols except 2-methylpropan-2-o1 gave carbonyl products: the primary alcohols gave aldehydes while the secondary and tertiary alcohols produced the corresponding ketones. With 0.5 mmol initial alcohol in 5 cm^3 water containing 50 mg TiO_2, these workers observed the evolution of H_2 in each system. The effectiveness of the TiO_2 with platinum black loading is much more than with platinum powder or ruthenium oxide. Bickley (2010) has showed the suitability of measuring photoactivity based on the zero-order conversion of 2-propanol to 2-propanone.

Recently, several workers have studied the partial conversion of alcohols to aldehydes. The partial oxidation of methanol to methanal, the simplest alcohol, can be achieved over TiO_2 with up to 91 % selectivity, without catalyst deactivation and product yields can be increased by elevating temperature up to 250 °C (Kominami et al. 2010). Similarly, selectivity values of the conversion of 2-propanol to acetone on platinum-containing TiO_2 solids were in the 78–80 % range at 22–28 % conversion (Lopez-Tenllado et al. 2012). Improvement in the photoactivity of TiO_2 for the conversion of 2-propanol (with 90–96 % selectivity), by synthesis on USY zeolite has been reported (Aramendia et al. 2007). Whereas single product is obtained with lower molecular weight alcohols, the oxidation of higher molecular weight alcohols may yield by-products. In a different study, the photocatalytic oxidation of cyclohexylmethanol, 2-cyclohexylethanol and 1-cyclohexylethanol over nano-P25 TiO_2 in acetonitrile under aerated conditions showed moderate conversion ratios of 55.2, 51.8 and 60.3 %, respectively, which comprise the corresponding aldehydes or ketones and carboxylic acids (Mohamed et al. 2008).

The photocatalytic oxidation of aqueous aryl alcohols to aldehydes at room temperature has been demonstrated by many workers. It is interesting to note that even heterocyclic alcohols such as furfuryl alcohol, undergo this type of photocatalytic reaction (Richard and Lamaire 1990). Selectivities are however relatively lower than those of aliphatic alcohols. Recently, the selective oxidation of benzyl alcohol to benzaldehyde in aqueous Cu(II)/TiO_2 suspension, under acidic conditions and solar UV irradiation was investigated by Marotta et al. (2011). At 72 % conversion of the initial substrate, 35 % benzaldehyde and 8 % benzoic acid were formed with respect to the initial benzyl alcohol concentration. By-products such as 2-hydroxy-benzyl-alcohol, 4-hydroxy-benzylalcohol, 2-hydroxy-benzaldehyde and 4-hydroxy-benzaldehyde were present at trace level which was attributed to the mediation of hydroxyl radicals. The possibility of improving typical benzaldehyde yields during photocatalytic synthesis has been demonstrated by Li et al. (2012). These workers achieved nearly 99 % yield and $\sim$ 100 % selectivity by using visible light irradiated TiO_2 nanorods.

Augugliaro et al. (2008) have reported the oxidation of aqueous benzyl alcohol and 4-methoxybenzyl alcohol in presence of suspensions of both laboratory prepared and commercial TiO_2 catalysts. For both alcohols the main oxidation products were the corresponding aromatic aldehydes and CO_2. The laboratory prepared TiO_2 catalysts exhibited selectivity values towards the aldehyde production up to 28 % with benzyl alcohol conversion reaching 50 % while the corresponding aldehyde yield was 41 % in the case of 4-methoxybenzyl alcohol with 65 % conversion, about four times higher than those of commercial TiO_2. The addition of an aliphatic alcohol (methanol, ethanol, 2-propanol or tert-butanol) in small amounts with respect to water decreased the overall oxidation rate of aromatic alcohols but enhanced the selectivity for aldehyde formation up to 1.5 times.

Surfactants may have great influence on the yield of products of benzyl alcohol oxidation. Bettoni et al. (2012) investigated the TiO_2 photosensitized oxidation of the water–insoluble 4-dodecyloxybenzyl alcohol in aqueous surfactant solutions.

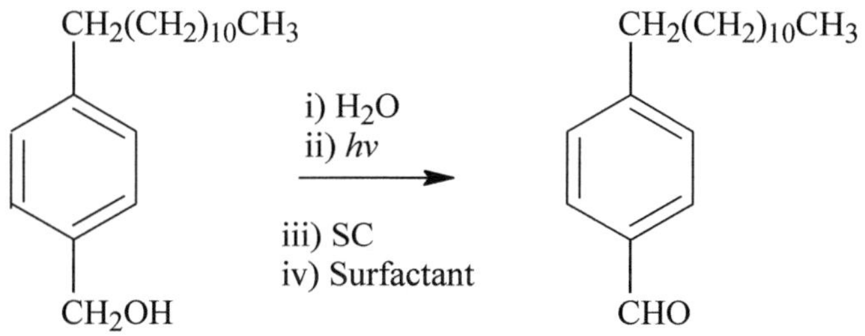

(6.10)

Surfactants used by these workers were of three different charge types as shown below.

Cationic surfactants:

1. Cetyltrimethylammonium chloride $CH_3(CH_2)_{14}CH_2N^+(CH_3)_3Cl$
2. Cetyltrimethylammonium bromide $CH_3(CH_2)_{14}CH_2N^+(CH_3)_3Br$
3. Methylcetyltrimethylammonium sulphite $CH_3(CH_2)_{14}CH_2N^+(CH_3)_3OSO_2CH_3$

Zwitterions:

1. Dimethyldodecylamine oxide, $CH_3(CH_2)_{12}CH_2N^+(CH_3)_2(CH_2)_3SO_3^-$
2. Tetradecyltrimethylammonium propane sulphonate, $CH_3(CH_2)_{14}CH_2N^+(CH_3)_3SO_3^-$
3. Tetradecyltripropylammonium propane sulphonate, $CH_3(CH_2)_{12}CH_2N^+(n\text{-}CH_3CH_2CH_3)_2(CH_2)_3SO_3^-$

Anionic surfactants:

1. Sodium dodecylsulphite, $CH_3(CH_2)_{12}CH_2OSO_3^-Na^+$
2. Sodium dodecylbenzenesulphonate, $CH_3(CH_2)_{12}CH_2ArSO_3^-Na^+$

Among these surfactants, dimethyldodecylamine oxide and cetyltrimethylammonium bromide showed the best performance. Dimethyldodecylamine oxide (0.015 moldm^{-3}) gave the best performance in water in terms of 4-dodecyloxybenzaldehyde yield (42 %) while cetyltrimethylammonium bromide in doses of 0.05 moldm^{-3} showed even higher performance (55 %) when acidified with 0.1 moldm^{-3} HCl. The mechanism suggested for the action of these surfactants was based on the co-micellisation of the benzylalcohol with the surfactant on the surface of the TiO_2 photocatalyst.

6.4 Photo-Epoxidation

Epoxides are chemical compounds of great importance as intermediate in the preparation of a wide variety of chemicals ranging from petrochemical compounds to fine chemicals such as perfumes, drugs, sweeteners and epoxy resins (Guo et al. 2011). These important intermediaries are commonly known to be highly reactive (Clayden et al. 2001). Since 1974, photosensitised epoxidation (by α–diketones for example) has been investigated (Suschitzky and Meth-Cohn 1980). Recently, the photocatalytic monoepoxidation of olefins has been investigated using TiO_2-pillared montmorillonite clay (Ouidri et al. 2010) (Eq. (6.11)). Cyclohexene epoxide was selectively produced as major product. The clay-pillared TiO_2 showed higher selectivity for cyclohexene epoxide than Degussa P25 TiO_2 due to the different specific surface area and hydrophobic nature of the pillared montmorillonite. After 8 h, the maximum yields of cyclohexene epoxide obtained with clay-pillared TiO_2 and P25 TiO_2 were 45 and 30 %, respectively, and these values tallied exactly with the product selectivites.

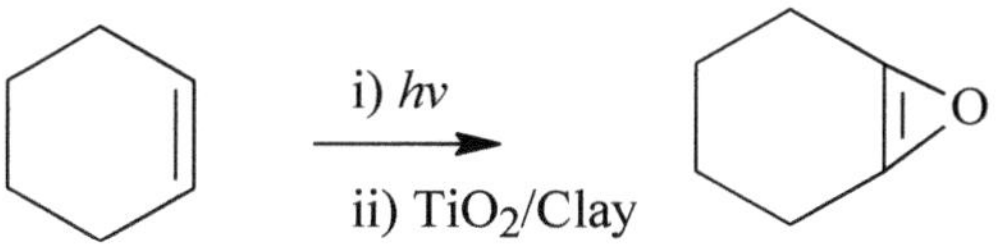

(6.11)

The hydroxylation of different ring substrates is common in photocatalysis. The hydroxylation of naphthalene to α-napthol in UV-irradiated TiO_2 suspensions has been reported by Huixian et al. (2011). The $^{\bullet}OH$ radicals generated on irradiation led to the direct hydroxylation of naphthalene to α-naphthol. The addition of $Fe^{2+} + H_2O_2$ remarkably increased the yield of α-naphthol from 22.2 % in bare-TiO_2. Even though surface-modified TiO_2 such as La-Eu/TiO_2, La-Y/TiO_2, $H_3PW_{12}O_{40}/TiO_2$, $H_3PMo_{12}O_{40}/TiO_2$, Fe/TiO_2, Ag/TiO_2, Cu/TiO_2, and N/TiO_2 can enhance the conversion and yield of α-naphthol, $TiO_2/Fe^{3+}/H_2O_2$ showed the highest yield (22.2 % at 89.5 % conversion).

From environmental viewpoint, the photocatalytic epoxidation of propene by molecular oxygen has been described as one of the best methods to produce propene oxide. The equation for such reaction of propene is given below.

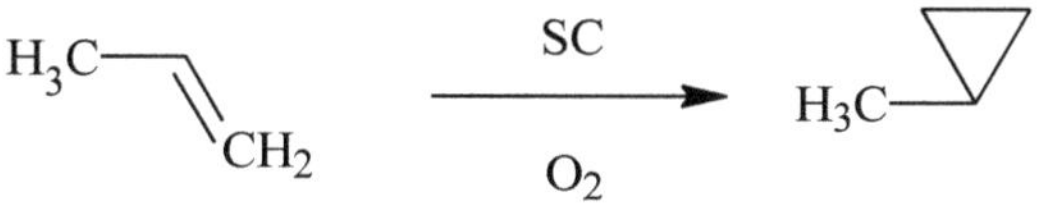

(6.12)

Nguyen et al. (2012) studied the gas-phase photocatalytic epoxidation of propene using a series of photocatalysts including SiO_2, TiO_2, V_2O_5/SiO_2 and Au/TiO_2, to evaluate their performance in photo-epoxidation. The gaseous mixture consisted of C_3H_6, O_2 and N_2 in the ratio 1:1:18 at atmospheric pressure. The by-products of the epoxidation reaction were propanaldehyde, acetone, acetaldehyde, ethanol and methanol. The semiconductor photocatalysts used in the study were outperformed by V-Ti/MCM-41 in terms of photo-oxidation rate (with 114×10^{-6} mol g^{-1}h-1) and selectivity (47 %) at 50 °C. This may be attributed to the higher recombination rate at the temperature range of the study (50–120 °C).

6.5 Routes to N-Containing Heterocyclics

Heterocyclic represent a large and important class of organic compounds. Dihydropyrazine is an intermediary of synthesis of an important heterocyclic compound pyrazine. This intermediate can be synthesized over zeolite (Hβ)-supported TiO_2 from equimolar slurry of ethylenediamine and propylene glycol in acetonitrile by a mechanism involving the photo-oxidation of the propylene glycol, then cyclization with ethylenediamine to give dihydromethylpyrazine, followed by subsequent oxidative demethylation forms dihydropyrazine (Rao et al. 2000). The dihydropyrazine yield was highest (20.4 %) in presence of 2 wt% TiO_2/Hβ zeolite. The steps involved in the mechanism are shown below.

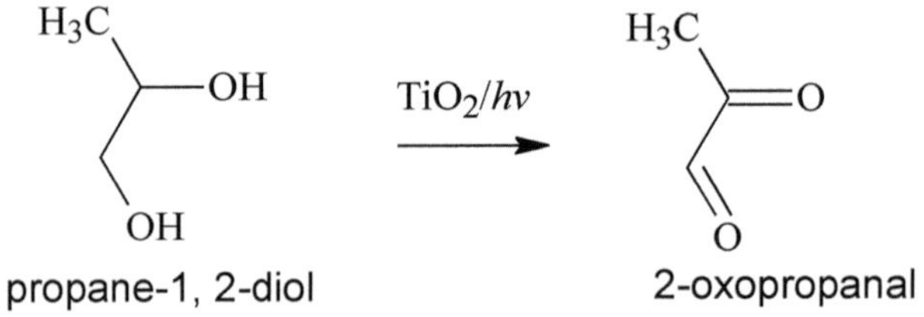

(6.13)

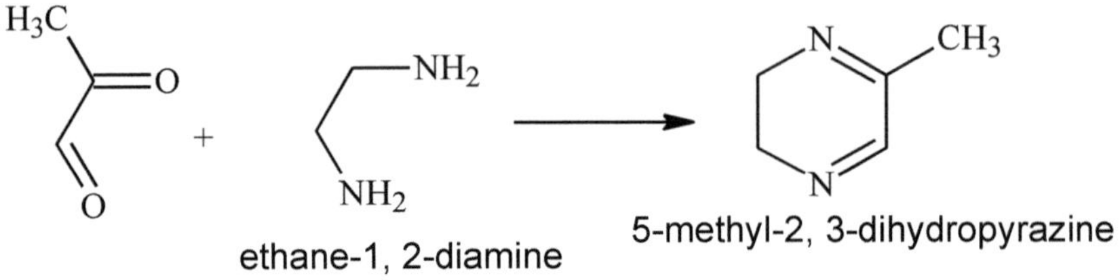

(6.14)

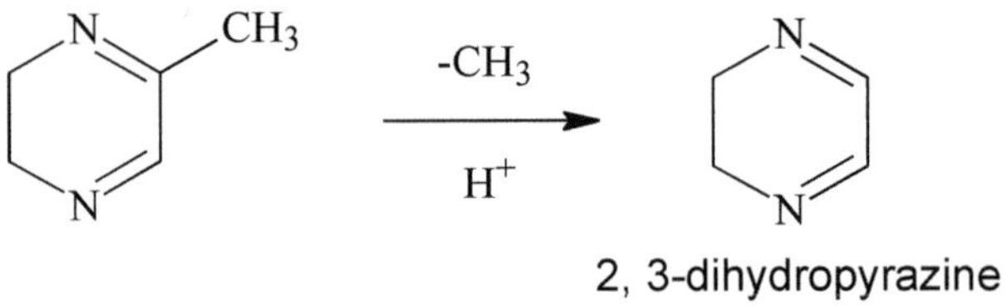

(6.15)

Similarly, piperazines have been formed by the irradiation of acetonitrile solution of N-(β-hydroxypropyl)ethylenediamine in presence of semiconductor–zeolite composite catalysts (Rao and Subrahmanyam 2002). The yield of 2-methylpiperazine and piperazine depends on the type of semiconductor and zeolite. For example, a 5 wt% TiO_2–Hβ zeolite composite showed the highest yield of 2-methylpiperazine (31.9 %).

Even though, C–C bond formation is the commonly attractive atom–atom coupling, other forms (such as C–N, C–O bond formation) are also of industrial interest. These processes can be in the form of condensation, addition, substitution etc. One of the oldest methods to facilely synthesise quinaldine, a heterocyclic base, is the Doebner–Miller method. This method involves the formation of crotonaldehyde via aldol condensation of acetaldehyde, and the latter reacts with aniline to yield 1,2-dihydro-1-methylquinoline which upon hydrogen loss gives quinaldine (Mann and Saunders 1978). The following steps are involved in the mechanism:
Aldol condensation:

$$2CH_3CHO \rightarrow CH_3CH(OH)CH_2CHO \tag{6.16}$$

Formation of crotonaldehyde:

$$CH_3CH\,(OH)\,CH_2CHO \rightarrow CH_3CH{=}CHCHO \tag{6.17}$$

Formation of intermediate:

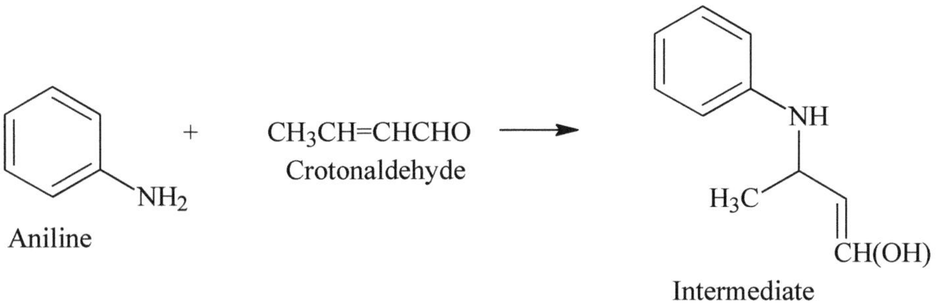

(6.18)

Cyclic dehydration of the intermediate:

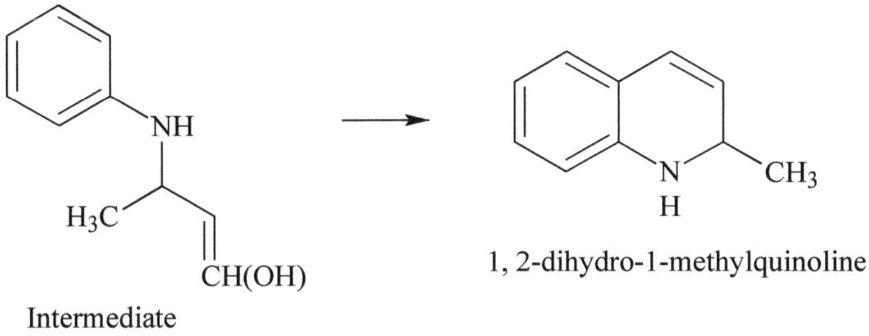

(6.19)

Hydrogen abstraction:

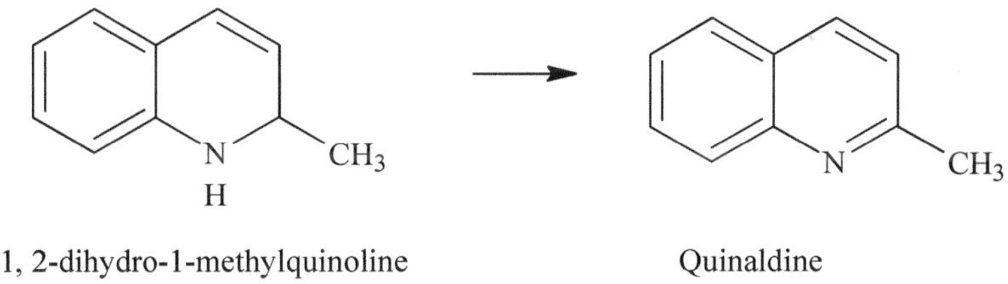

(6.20)

Even though the Doebner-Miller method of synthesis of quinaldine easily produces the crude compound, it entails the need to remove several by-products such as secondary amines. Recently, the synthesis of quinaldine from different nitrogen-containing arenes has been successfully achieved by photocatalytic methods. Selvam and Swaminathan (2012) reported a one-pot synthesis of quinaldine and its derivatives from nitrobenzene and ethanol over UV/visible light illuminated N-TiO_2. The mechanism described by these workers was similar to the Doebner-Miller type in that condensation-cyclisation takes place, but a distinguishing fact is that the photocatalytic method of synthesis involves light illumination rather than heating. In the one-pot synthesis under visible light, N-doped TiO_2 was found to be more efficient than metal doped TiO_2 which confirms spectral dependence of the reaction.

Prior to this study, the same workers demonstrated the photocatalytic synthesis of quinaldines from different anilines in ethanol in the presence of Au-loaded TiO_2 (Selvam and Swaminathan 2010). Notably, aniline conversion up to 90 % could be achieved while higher conversion efficiency (94 %) was obtained when *p*-phenatidine was used as the starting material. One of the plausible mechanisms suggested for the photoconversion of ethanolic aniline to quinaldine over Au-TiO_2 is shown below:
Formation of crotonaldehyde:

$$CH_3CH_2OH + CH_3CHO \rightarrow CH_3CH{=}CHCHO \qquad (6.21)$$

Formation of intermediate:

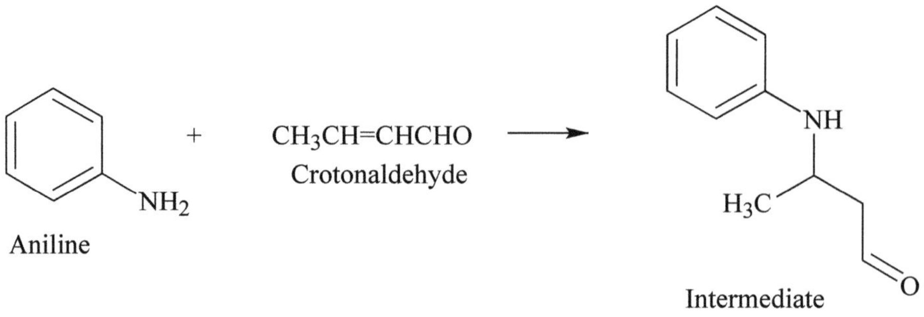

(6.22)

Cyclisation of the intermediate:

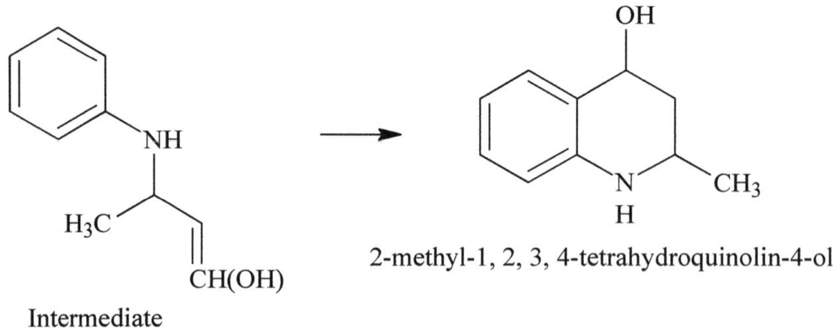

(6.23)

Hydrogen abstraction:

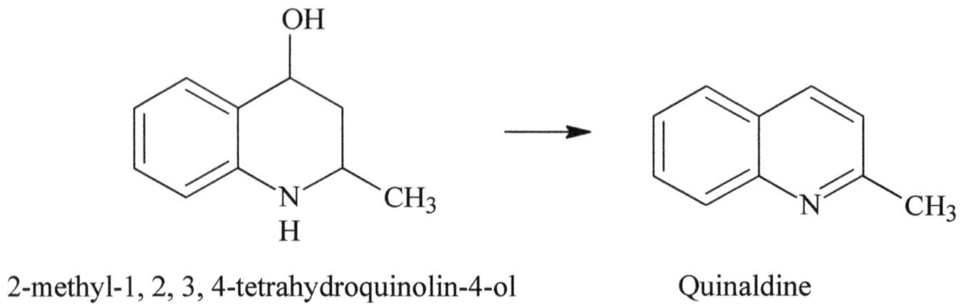

(6.24)

It would be noted that the structure of the cyclisation product of the photocatalytic conversion differs from that of the thermal Doebner-Miller conversion. This may be attributed to the fact that different starting materials were used and that photo-processes usually lead different reaction mechanisms from thermal reactions. With silver doping, aniline conversions in presence of ethanol can reach 99 % without any change in the mechanism of the photoconversion (Selvam and Swaminathan 2011a).

Various alkyl and aryl aldehydes including heterocyclic forms can be used to synthesize *o*-substituted benzothiazoles. The synthesis was recently demonstrated with significant yield for the first time by Das et al. (2013) using nanospherical CdS particles under visible light irradiation ($\lambda = 420$ nm). The yields of products were solvent dependent. Ethanol, chloroform and methanol are good solvents for the conversion of substituted benzothiazoles with 60, 90 and 96 % yields. The general equations for this conversion are shown below:

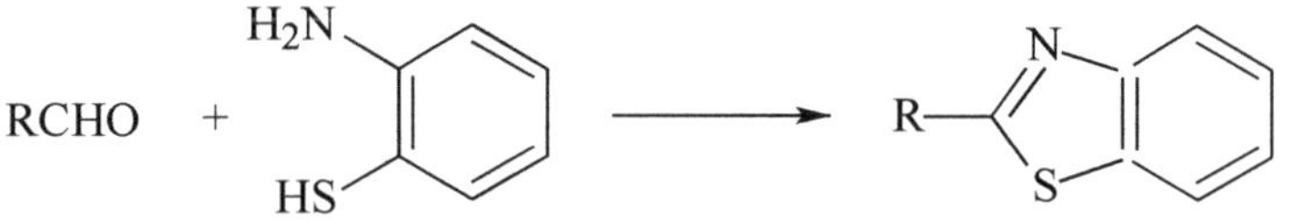

2-aminobenzenethiol 2-alkyl-4, 5-dihydro-1, 3-thiazole-benzene (1:1)

(6.25)

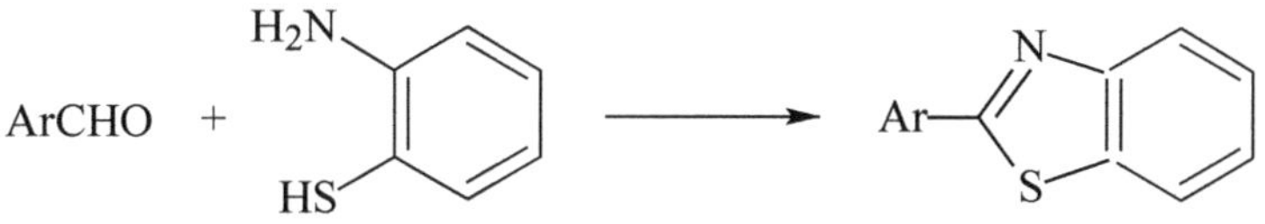

2-aminobenzenethiol 2-aryl-4, 5-dihydro-1, 3-thiazole-benzene (1:1)

(6.26)

where Ar represents substituted or unsubstituted phenyl rings and R represents an alkyl group.

Over the years, the radical addition of tertiary amines to alkenes has been a well-known reaction. The photoinduced radical addition of N-methylpyrrolidine to various electron-deficient α,β-unsaturated lactones using TiO_2 as photosensitiser has been reported by Hoffmann et al. (2006). A representative example is tthe synthesis of methyl pyrrolidine substituted furanone given by Eq. (6.27). These reactions are clear examples of photocatalytic C-C coupling.

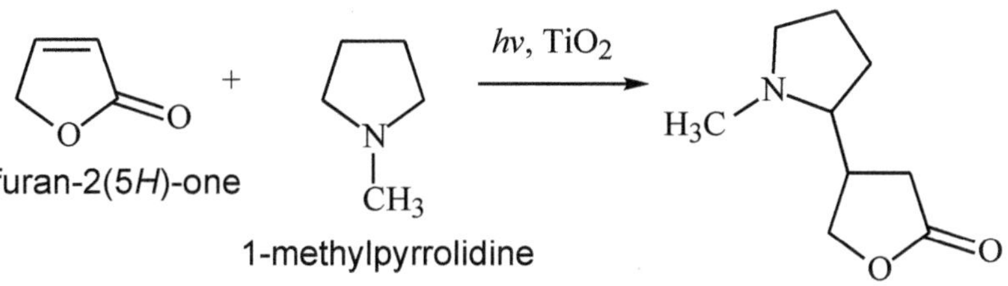

furan-2(5*H*)-one

1-methylpyrrolidine

4-(1-methylpyrrolidin-2-yl)dihydrofuran-2(3*H*)-one

(6.27)

One-pot synthesis of disubstituted benzimidazoles from N-substituted alcoholic 2-nitroanilines or 1,2-diamines with 3–12 nm-sized platinum particles loaded on the TiO_2 using solar and UV-A light is described (Selvam and Swaminathan 2011b). The mechanism suggested for the reaction is based on the following sequence of steps:

1. electron excitation from the valence band of TiO_2 to the conduction band

$$TiO_2 \rightarrow (e^-_{CB}, h^+_{VB}) \quad (6.28)$$

2. transfer of the conduction band electrons to Pt
3. synchronous reduction of the alcohol and and oxidation of the N-substituted 2-nitroanilines in presence of TiO_2 (oxidising agent) and Pt (reducing agent) to form N-substituted diamines.

Reduction of alcohol:

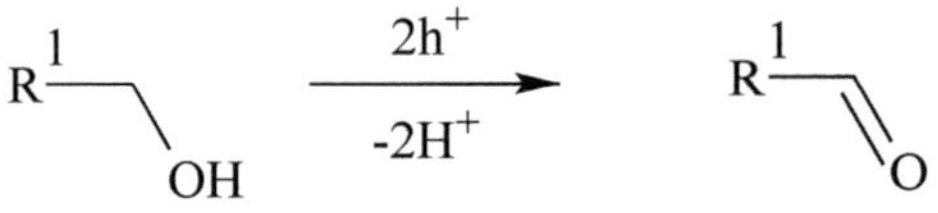

R1 = Alkyl group (6.29)

Oxidation of substituted nitroaniline:

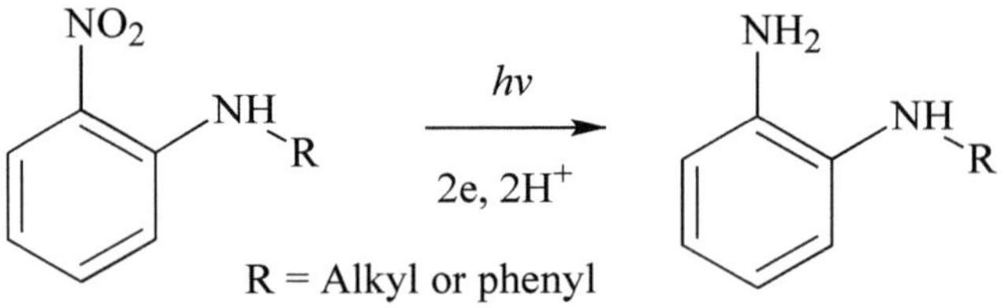

R = Alkyl or phenyl (6.30)

4. reductive cyclisation of the N-substituted diamines

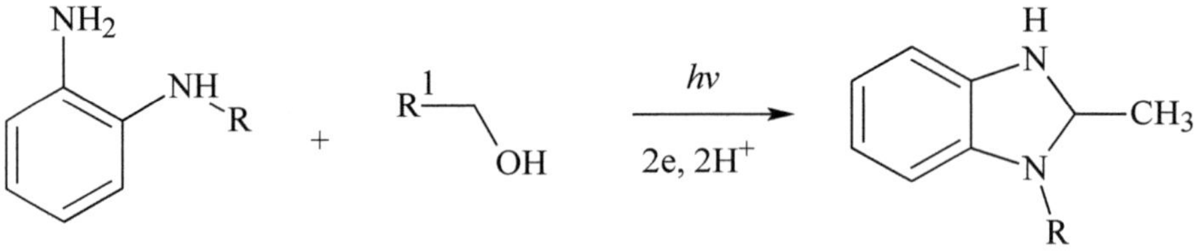

(6.31)

5. Oxidation of N-substituted diamines to disubstituted benzimidazoline

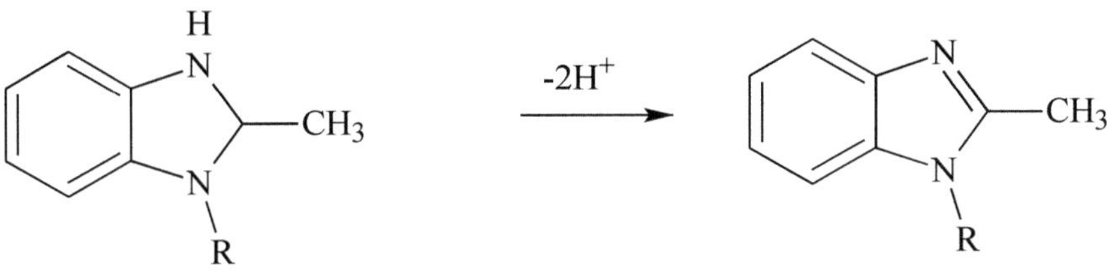

(6.32)

Various alcohols and N-substituted 2-nitroanilines can undergo this type of reaction to yield the disubstituted benzimidazole. The H^+ produced in the process is believed to be reduced by trapped electrons on the surface of Pt.

6.6 Synthesis of Amines

Alkanes and alkenes constitute a large percentage of hydrocarbon by-products in the petroleum and petrochemical industries, often in the form of gases. Semiconductor photocatalysis is a proven candidate for the selective conversion of these by-products to useful products such as amines. The photocatalytic addition of cyclopentene and cyclohexene to immines to yield saturated amines has been investigated by Shindler and Kisch (1997) using methanolic CdS suspensions as catalyst. The immines in either the case of cyclopentene or cyclohexene gave the corresponding ammines as major products with a hydrodimer of the immine as by-product as show in Eq. (6.33) and (6.34).

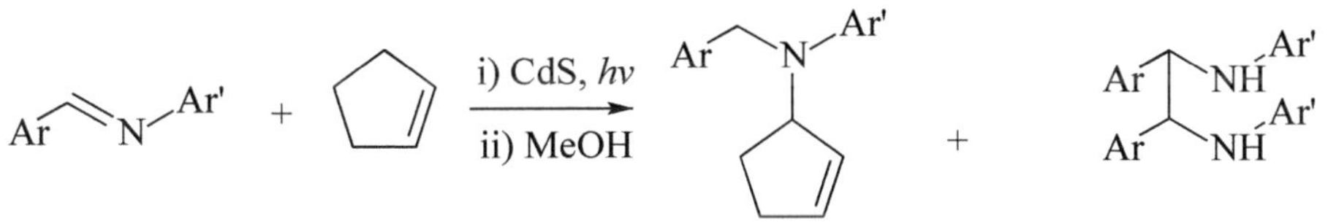

(6.33)

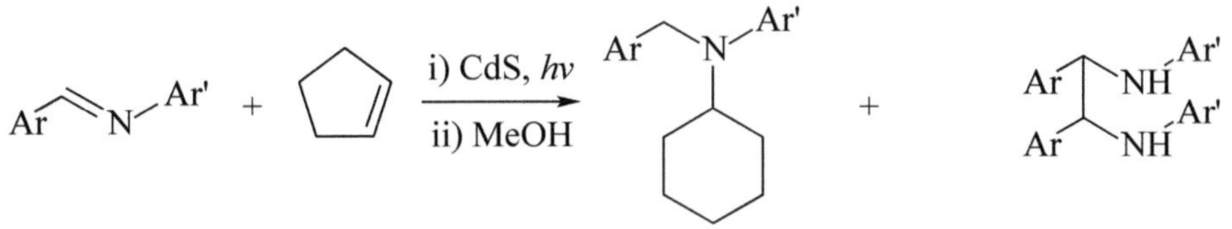

(6.34)

The deactivation of semiconductor in the process can be avoided by humidification of the air stream (Carneiro et al. 2011).

6.7 Photo-Hydroxylation

Photocatalytic hydroxylation can directly give rise to economically important aromatic compounds. In the direct photocatalytic synthesis of α-naphthol from naphthalene, •OH radicals generated on UV-illuminated TiO_2 photocatalysts led to the direct hydroxylation of naphthalene to α-naphthol (Huixian et al. 2011). Inorganic systems such as Fe^{3+}, Fe^{2+}, $Fe^{3+} + H_2O_2$, and $Fe^{2+} + H_2O_2$ greatly enhanced the conversion of naphthalene and the yield of α-naphthol in the semiconductor suspensions. The addition of $Fe^{3+} + H_2O_2$ to a TiO_2 suspension increased the yield to 22.2 %. The hydroxylation reaction was also significantly enhanced by surface modification of the TiO_2 to La-Eu/TiO_2, La-Y/TiO_2, $H_3PW_{12}O_{40}$/TiO_2, $H_3PMo_{12}O_{40}$/TiO_2, Fe/TiO_2, Ag/TiO_2, Cu/TiO_2, N/TiO_2 and especially Fe/TiO_2.

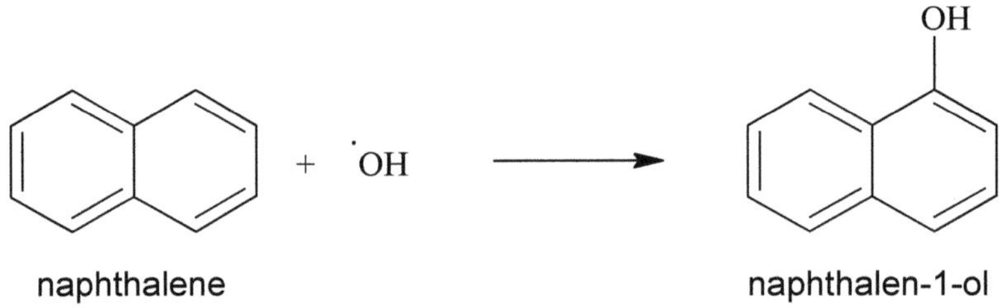

(6.35)

The direct photocatalytic hydroxylation of benzene to phenol in presence of semiconductor particles is a potentially attractive route to industry as an alternative of the cumene process and this process has been established with remarkable turnover numbers (Zama et al. 2000). The properties of the semiconductor photocatalyst can influence the regioselectivity and the distribution of regio-isomers formed by photocatalytic hydroxylation reaction. Bellardita et al. (2012) detected ortho- and para- monohydroxyphenol as the the products of the primary oxidation of phenol while those of benzoic acid were all the mono-hydroxy derivatives. The ortho-isomer was not detected in the presence of home prepared TiO_2 catalysts used at substrate initial concentration lower than 1×10^{-3} moldm^{-3} (as opposed to the para-one) owing of the strong interaction of ortho-monohydroxy derivative with catalyst surface. The effect of substituent groups on selectivity to hydroxylated aromatics was investigated in details by Palmisano et al. (2007) using substrates containing a deactivating group (nitrobenzene, cyanobenzene, benzoic acid, 1-phenylethanone), an activating group (phenol, phenylamine, N-phenylacetamide) or both a deactivating group and activating group (4-chlorophenol). The results revealed that the primary photocatalytic oxidation of aromatic compounds containing an electron donating group yielded mainly ortho and para mono-hydroxy derivatives while in the presence of an electron withdrawing group all the mono-hydroxy derivatives are obtained.

6.8 Other Routes of Organic Synthesis

One of the important ingredients used in foods, drugs, cosmetics and fine chemicals manufacture is vanillin. This important compound has been successfully synthesized by Augugliaro et al. (2012) from trans-ferulic acid (**1**), isoeugenol (**2**), eugenol (**3**) and vanillyl alcohol (**4**) in aqueous TiO_2 suspensions. The photosynthesis of vanillin had selectivity ranging from 1.4 to 21 mol % with respect to the converted substrate.

(1)
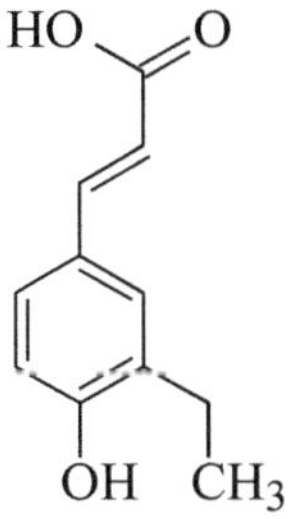

Ferulic acid

(2)
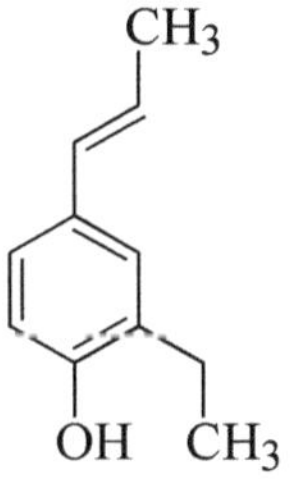

Isoeuginol

(3)
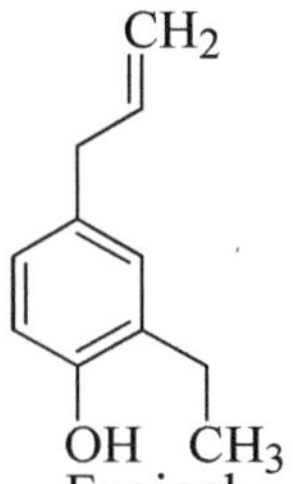

Euginol

(4)
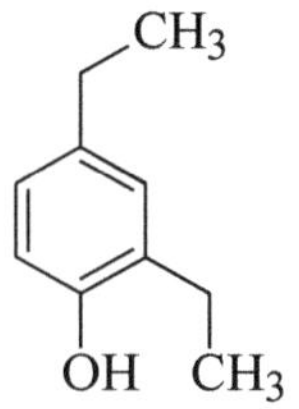

Vanillyl alcohol

The intermediates of trans-ferulic acid conversion to vanillin (**5**) were homovanillic acid (**6**), vanillyl mandelic acid (**7**) with some amount of trans-caffeic acid (**8**). The mechanism for intermediate formation was based on the action of hydroxyl radicals.

(5)
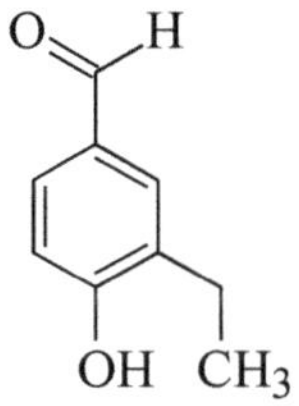

Vanillin

(6)
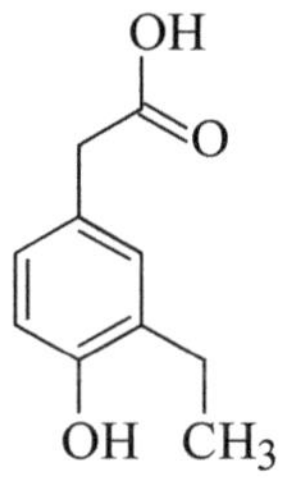

Homovanillic acid

(7)
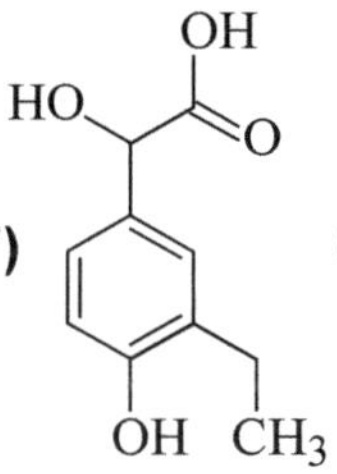

Vanillyl mandelic acid

(8)
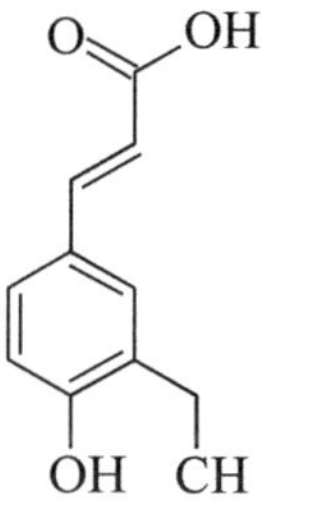

Trans-caffeic acid

6.9 Conclusion

The potentialities of photocatalysis in organic synthesis can be clearly seen from the preceeding highlights. Photocatalytic synthesis would lessen cost of operation because reactions take place at ambient conditions. Other important attributes of photocatalytic synthesis include eco-friendly nature and easy monitoring of the processes. An interesting fact, which we may deduce from the synthesis of quinaldines highlighted in this chapter, is that not all the starting materials in conventional synthesis may be required for photocatalytic synthesis. In addition, most of the synthesis methods based on semiconductor photocatalysis require shorter time typically 30 to

120 min. The living challenge is how to scale-up these emerging methods for potential harvesting at large scale.

References

Aramendía MA, Colmenares JC, López-Fernández S et al (2007) Screening of different zeolite-based catalysts for gas-phase selective photooxidation of propan-2-ol. Catal Today 129:102–109

Augugliaro V, Camera-Roda G, Loddo V et al (2012) Synthesis of vanillin in water by TiO_2 photocatalysis. Appl Catal A 111–112:555–561

Augugliaro V, Kisch H, Loddo V et al (2008) Photocatalytic oxidation of aromatic alcohols to aldehydes in aqueous suspension of home-prepared titanium dioxide 1. Selectivity enhancement by aliphatic alcohols. Appl Catal A 349:182–188

Bellardita M, Augugliaro V, Loddo V et al (2012) Selective oxidation of phenol and benzoic acid in water via home-prepared TiO_2 photocatalysts: distribution of hydroxylation products. Appl Catal A 441–442:79–89

Bettoni M, Brinchi L, Giacco TD et al (2012) Surfactant effect on titanium dioxide photosensitized oxidation of 4-dodecyloxybenzyl alcohol. J Photochem Photobiol A 229:53–59

Bickley RI (2010) Heterogeneous photocatalysis at liquid–solid interfaces. Oxidative dehydrogenation of propan-2-ol as a method of assessing photocatalytic activity? J Photochem Photobiol A 216:256–260

Carneiro JT, Yang C-C, Moulijn JA et al (2011) The effect of water on the performance of TiO_2 in photocatalytic selective alkane oxidation. J Catal 277:129–133

Clayden J, Warren S, Greeves N et al (2001) Organic chemistry. Oxford University Press, New York

Das S, Samanta S, Maji SK et al (2013) Visible-light-driven synthesis of 2-substituted benzothiazoles using CdS nanosphere as heterogenous recyclable catalyst. Tetrahedron Lett 54:1090–1096

Ferry JL, Glaze WH (1998) Photocatalytic reduction of nitroorganics over illuminated titanium dioxide: Electron transfer between excited-state TiO_2 and nitroaromatics. J Phys Chem B 102:2239–2244

Fox MA, Dulay MT (1995) Heterogeneous Photocatalysis. Chem Rev 93:341–357

Guo Y, Liu Z, Wang G et al (2011) Highly selective epoxidation of styrene over mesoporous Au–Ti–SBA-15 via photocatalysis process: Synthesis, characterization, and catalytic application. Appl Surf Sci 258:1082–1090

Hoffmann N, Bertrand S, Marinković S et al (2006) Efficient radical addition of tertiary amines to alkenes using photochemical electron transfer. Pure Appl Chem 78:2227–2246

Huixian S, Tianyong Z, Hongliang W et al (2011) Photocatalytic conversion of naphthalene to α-naphthol using nanometer-sized TiO_2. Chin J Catal 32:46–50

Jackson SD, Hargreaves JSJ, Lennon D (2003) Catalysis in application. Royal Society of Chemistry, Athenaeum Press Ltd, UK

Kominami H, Sugahara H, Hashimoto K (2010) Photocatalytic selective oxidation of methanol to methyl formate in gas phase over titanium(IV) oxide in a flow-type reactor. Catal Commun 11:426–429

Kraeutler B, Bard AJ (1978) Heterogeneous photocatalytic decomposition of saturated carboxylic acids on TiO_2 powder. Decarboxylative route to alkanes. J Am Chem Soc 100:5985–992

Li C-J, Xu G-R, Zhang B et al (2012) High selectivity in visible-light-driven partial photocatalytic oxidation of benzyl alcohol into benzaldehyde over single-crystalline rutile TiO_2 nanorods. Appl Catal B 115–116:201–208

Lopez-Tenllado FJ, Marinas A, Urbano FJ et al (2012) Selective photooxidation of alcohols as test reaction for photocatalytic activity. Appl Catal B 128:150–158

Mann FG, Saunders BC (1978) Practical organic chemistry. Longman, New York

Marotta R, Di Somma I, Spasiano D et al (2011) Selective oxidation of benzyl alcohol to benzaldehyde in water by TiO_2/Cu(II)/UV solar system. Chem Eng J 172:243–249

Mohamed OS, Ahmed SA, Mostafa MF et al (2008) Nanoparticles TiO_2-photocatalyzed oxidation of selected cyclohexyl alcohols. J Photochem Photobiol A 200:209–215

Nguyen V-H, Chan H-Y, Wu JCS et al (2012) Direct gas-phase photocatalytic epoxidation of propylene with molecular oxygen by photocatalysts. Chem Eng J 179:285–294

Nishimoto S-I, Ohtani B, Kagiya T (1985) Photocatalytic dehydrogenation of aliphatic alcohols by aqueous suspensions of platinized titanium dioxide. J Chem Soc Faraday Trans 1(81):2467–2474

Ouidri S, Guillard C, Caps V et al (2010) Epoxidation of olefins on photoirradiated TiO_2-pillared clays. Appl Clay Sci 48:431–437

Palmisano G, Addamo M, Augugliaro V et al (2007) Selectivity of hydroxyl radical in the partial oxidation of aromatic compounds in heterogeneous photocatalysis. Catal Today 122:118–127

Rao KBS, Srinivas B, Prasad AR et al (2000) A novel one step photocatalytic synthesis of dihydropyrazine from ethylenediamine and propylene glycol. Chem Commun 16:1533–1534

Rao KVS, Subrahmanyam M (2002) Synthesis of 2-methylpiperazine by photocatalytic reaction in a non-aqueous suspension of semiconductor–zeolite composite catalysts. Photochem Photobiol Sci 1:597–599

Richard C, Lemaire J (1990) Analytical and kinetic study of the phototransformation of furfuryl alcohol in aqueous ZnO suspensions. J Photochem Photobiol A 55:127–134

Sato S (1982) Photocatalytic synthesis of ethane from acetic acid. Kolbe reaction at a gas-solid interface under U.V.-illumination. J Chem Soc Chem Commun 1:26–27

Selvam K, Swaminathan M (2010) Au-doped TiO_2 nanoparticles for selective photocatalytic synthesis of quinaldines from anilines in ethanol. Tetrahedron Lett 51:4911–4914

Selvam K, Swaminathan M (2011a) Cost effective one-pot photocatalytic synthesis of quinaldines from nitroarenes by silver loaded TiO_2. J Mol Catal A: Chem 351:52–61

Selvam K, Swaminathan M (2011b) An easy one-step photocatalytic synthesis of 1-aryl-2-alkylbenzimidazoles by platinum loaded TiO_2 nanoparticles under UV and solar light. Tetrahedron Lett 52:3386–3392

Selvam K, Swaminathan M (2012) Nano N-TiO_2 mediated selective photocatalytic synthesis of quinaldines from nitrobenzenes. RSC Adv. doi:10.1039/c2ra01178f

Shindler W, Kisch H (1997) Heterogeneous photocatalysis XV. Mechanistic aspects of cadmium sulphide-catalyzed photoaddition of olefins to Schiff bases. J Photochem Photobiol A 103:257–264

Suschitzky H, Meth-Cohn O (1980) Heterocyclic chemistry, Special periodical reports, vol 1. Royal Society of Chemistry, London

Wu W, Wen L, Shen L et al (2013) A new insight into the photocatalytic reduction of 4-nitroaniline to p-phenylenediamine in the presence of alcohols. Appl Catal, B 130–131:163–167

Zama K, Fukuoka A, Sasaki Y et al (2000) Selective hydroxylation of benzene to phenol by photocatalysis of molybdenum complexes grafted on mesoporous FSM-16. Catal Lett 66:251–253

Index

U. I. Gaya, *Heterogeneous Photocatalysis Using Inorganic Semiconductor Solids*, DOI 10.1007/978-94-007-7775-0,

MIX
Papier aus verantwortungsvollen Quellen
Paper from responsible sources
FSC® C105338

If you have any concerns about our products,
you can contact us on
ProductSafety@springernature.com

In case Publisher is established outside the EU,
the EU authorized representative is:
Springer Nature Customer Service Center GmbH
Europaplatz 3, 69115 Heidelberg, Germany

Printed by Libri Plureos GmbH
in Hamburg, Germany